Collins

ROYAL
OBSERVATORY
GREENWICH

Night Sky
ALMANAC

A STARGAZER'S
GUIDE TO

2023

Storm Dunlop & Wil Tirion

Published by Collins
An imprint of HarperCollins Publishers
Westerhill Road
Bishopbriggs
Glasgow G64 2QT
www.harpercollins.co.uk

In association with
Royal Museums Greenwich, the group name for the National Maritime Museum,
Royal Observatory Greenwich, the Queen's House and *Cutty Sark*
www.rmg.co.uk

The contents of this publication are believed correct at the time of printing.
Nevertheless the publisher can accept no responsibility for errors or omissions,
changes in the detail given or for any expense or loss thereby caused.

HarperCollins does not warrant that any website mentioned in this title will
be provided uninterrupted, that any website will be error free, that defects will
be corrected, or that the website or the server that makes it available are free
of viruses or bugs. For full terms and conditions please refer to the site
terms provided on the website.

A catalogue record for this book is available from the British Library

ISBN 978-0-00-853259-8

10 9 8 7 6 5 4 3 2 1

Printed in the UK using 100% Renewable Electricity at
CPI Group (UK) Ltd

If you would like to comment on any aspect of this book,
please contact us at the above address or online.
e-mail: collins.reference@harpercollins.co.uk

facebook.com/CollinsAstronomy
@CollinsAstro

This book is produced from independently certified FSC™ paper
to ensure responsible forest management.

For more information visit: www.harpercollins.co.uk/green

Contents

Introduction

The aim of this book is to help people to find their way around the night sky and to understand what is visible every month, from anywhere in the world. The stars that may be seen depend on where you are on Earth, but even if you travel widely, this book will show you what you can see. The night sky also changes from month to month and these changes, together with some of the significant events that occur during the year are described and illustrated.

The charts that are used differ considerably from those found in most astronomy books, and have been specifically designed for use anywhere in the world. A full description of how to use and understand the monthly charts is given on pages 36–39.

Sunrise, sunset and twilight
The conditions for observing naturally vary over the course of the year and one's location on Earth. Sunrise and sunset vary considerably, depending in particular on one's latitude. Sunrise and sunset times are given each month for nine different locations around the world. These places are shown in a **bold** typeface on the world map on pages 40–41. Sunrise and sunset times are given for the first and last days in every month, for these specific locations. Another factor that influences what may be seen is twilight at dusk and dawn. Again, this varies considerably with one's latitude on Earth. The diagrams on pages 251–253 show how this varies for the nine locations, which have been chosen to show the range of variation, rather than just for the importance of the places that have been included. The different stages of twilight and how they affect observing are also explained there.

Moonlight
Yet another factor that affects the visibility of objects is the amount of moonlight in the sky. At Full Moon, it may be very difficult to see some of the fainter stars and objects, and even when the Moon is at a smaller phase it seriously interferes with visibility if it is near the stars or planets in which you are interested. A full lunar calendar is given for each month and

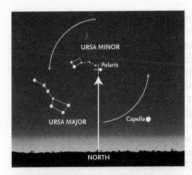

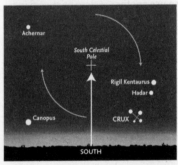

The altitude of the Celestial Pole equals the observer's latitude.

may be used to see when nights are likely to be darkest and best for observing.

The celestial sphere
All the objects in the sky (including the Sun, Moon, and stars) appear to lie at some indeterminate distance on a large sphere, centred on the Earth. This celestial sphere has various reference points and features that are related to those of the Earth. If the Earth's rotational axis is extended, for example, it points to the North and South Celestial Poles, which are thus in line with the North and South Poles on Earth. As shown in the diagrams, the altitude of the celestial pole is equal to the observer's latitude, whether in the north or south. Similarly, the celestial equator lies in the same plane as the Earth's equator, and divides the sky into northern and southern hemispheres.

It is useful to know some of the special terms for various parts of the sky. As seen by an observer, half of the celestial sphere is invisible, below the horizon. The point directly overhead is known as the *zenith*, and this point is shown on the monthly charts for several different latitudes, where it is an important reference point. The (invisible) point below one's feet is the **nadir**. The line running from the north point on the horizon, up through the zenith and then down to the south point is the **meridian**. This is an important invisible line in the sky, because objects are highest in the sky, and thus easiest to see, when they cross the meridian in the south. Objects are said to transit, when they cross this line in the sky.

In this book, reference is sometimes made in the text and in the diagrams to the standard compass points around the horizon. The position of any object in the sky may be described by its **altitude** (measured in degrees above the horizon) and its **azimuth** (measured in degrees from north, 0°, through east, 90°, south, 180°, and west, 270°). Experienced amateurs and professional astronomers also use another system of specifying locations on the celestial sphere, but that need not concern us here, where the simpler method will suffice.

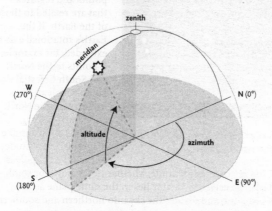

Measuring altitude and azimuth on the celestial sphere.

The celestial sphere appears to rotate about an invisible axis, running between the north and south celestial poles. The location (i.e., the altitude) of the celestial poles depends entirely on the observer's position on Earth or, more specifically, their latitude.

Right ascension and declination

On the previous pages we have mentioned how a simple method, involving altitude and azimuth, measured relative to the observer's horizon, may be used to specify the position of an object in the sky. Astronomers, however, use another, more precise method, which does not depend on the observer's position on Earth (and thus on their local horizon). This involves the two co-ordinates, *right ascension* (RA) and *declination* (dec). Right ascension is measured eastwards (to the left) from the *First Point of Aries* (page 78) in hours and minutes of time (and very occasionally, in seconds) or else (less frequently) in degrees. Because the Earth rotates once in 24 hours, one hour of right ascension equals 15 angular degrees. The sky appears to rotate by this amount in one hour.

All objects in the sky appear to be located on an imaginary sphere: the celestial sphere. There are, however, certain fixed points on the celestial sphere, related to points on the Earth. The North Celestial Pole (NCP) and the South Celestial Pole (SCP) are located in line with the projection of the Earth's rotational axis onto that sphere. In the north, the NCP is very close to Polaris, which has been known as the North Star since antiquity. In a similar way, the celestial equator is the projection onto the sphere of the Earth's equator. The second co-ordinate, declination, is simply the angular distance, in degrees, north or south of the celestial equator. The Sun has a declination of zero when it appears to cross the celestial equator at the equinoxes.

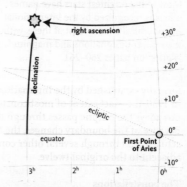

Measuring right ascension (RA) and declination (dec) on the celestial sphere.

The ecliptic and the zodiac

Another important line on the celestial sphere is the Sun's apparent path against the background stars – in reality the result of the Earth's orbit around the Sun. This is known as the *ecliptic*. The point where the Sun, apparently moving along the ecliptic, crosses the celestial equator from south to north is known as the vernal (or northern spring) equinox, which occurs around March 21. At this time (and at the northern autumnal equinox, on September 22 or 23, when the Sun crosses the celestial equator from north to south) day and night are almost exactly equal in length. (There is a slight difference, but that need not concern us here.) The vernal equinox is currently located in the constellation of Pisces, and is important in astronomy because it defines the zero point for a system of celestial coordinates, which is, however, not used in this book.

The Moon and planets are to be found in a band of sky that extends 8° on either side of the ecliptic. This is because the orbits of the Moon and planets are inclined at various angles to the ecliptic (i.e., to the plane of the Earth's orbit). This band of sky is known as the zodiac, and when originally devised, consisted of twelve *constellations*, all of which were considered to be exactly 30° wide. When the constellation boundaries were formally established by the International Astronomical Union in 1930, the exact extent of most constellations was altered, and nowadays the ecliptic passes through thirteen constellations. Because of the boundary changes, the Moon and planets may actually pass through several other constellations that are adjacent to the original twelve.

Most of the brightest stars have names officially recognized by the International Astronomical Union. A list of these, with their Bayer designations and magnitudes, is given on pages 260–261.

The constellations

In the western astronomical tradition, the celestial sphere has always been divided into various constellations, most dating back to antiquity and usually associated with certain myths or legendary people and animals. Nowadays, 88 constellations cover the whole sky, and their boundaries have been fixed by international agreement. Their names (in Latin) are largely derived from Greek or Roman originals. (A full list of the constellations is given on pages 256–258, with their names in English, their abbreviations, and genitive forms.) Some of the names of the most prominent stars are of Greek or Roman origin, but many are derived from Arabic names. Some bright stars have no individual names, and for many years, they were identified by terms such as 'the star in Hercules' right foot'. A more sensible scheme was introduced by the German astronomer Johannes Bayer in the early seventeenth century. Following his scheme – which is still used today – most of the brightest stars are identified by a Greek letter followed by the genitive form of the constellation's Latin name. An example is the Pole Star, also known as Polaris and α Ursae Minoris. The Greek alphabet is shown on page 258.

Asterisms

Apart from the constellations, certain groups of stars, which may form a small part of a larger constellation, are readily recognizable and have been given individual names. These groups are known as asterisms, and the most famous (and well-known) is the 'Plough' or 'Big Dipper', the common name for the seven brightest stars in the constellation of Ursa Major, the Great Bear. The names and identifications of some popular asterisms are given in the list on page 259.

The Moon

As it passes across the sky from west to east in its orbit around the Earth, the Moon moves by approximately its diameter (about half a degree) in an hour. Normally, in its orbit, the Moon passes above or below the direct line between Earth and Sun (at New Moon) or outside the area obscured by the Earth's shadow (at Full Moon). Occasionally, however, the three bodies are more-or-less perfectly aligned to give an eclipse: a solar eclipse at New Moon, or a lunar eclipse at Full Moon. Depending on

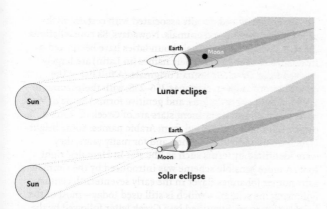

When the Moon passes through the Earth's shadow (top), a lunar eclipse occurs. When it passes in front of the Sun (below) a solar eclipse occurs.

the exact circumstances, a solar eclipse may be merely partial (when the Moon does not cover the whole of the Sun's disc); annular (when the Moon is too far from Earth in its orbit to appear large enough to hide the whole of the Sun); or total. Total and annular eclipses are visible from very restricted areas of the Earth, but partial eclipses are normally visible over a wider area. Two forms of solar eclipse occur this year, and are described in detail in the appropriate month.

Precautions must always be taken when viewing even partial phases of a solar eclipse to avoid damage to your eyes. Only ever use proper eclipse glasses, or a proper solar filter over the full objective of a telescope. The glass 'solar filters' sometimes provided with cheap telescopes should never be used. They are unsafe.

Somewhat similarly, at a lunar eclipse, the Moon may pass through the outer zone of the Earth's shadow, the penumbra (in a penumbral eclipse, which is not generally perceptible to the naked eye); pass so that just part of the Moon is within the darkest part of the Earth's shadow, the umbra (in a partial eclipse); or completely within the umbra (in a total eclipse).

11

Unlike solar eclipses, lunar eclipses are visible from large areas of the Earth. Again, these are described in detail in the relevant month.

Occasionally, as it moves across the sky, the Moon passes between the Earth and individual planets or distant stars, giving rise to an *occultation*. As with solar eclipses, such occultations are visible from restricted areas of the world, but certain significant occultations are described in detail.

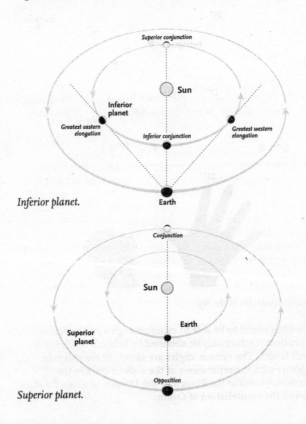

Inferior planet.

Superior planet.

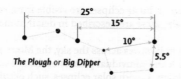

The Plough or Big Dipper

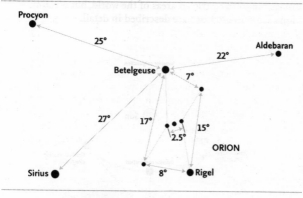

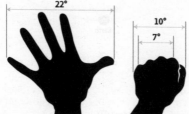

Measuring angles in the sky.

It is often useful to be able to estimate angles on the sky, and approximate values may be obtained by holding one hand at arm's length. The various angles are shown in the diagram, together with the separations of the various stars in the asterism, known as the Plough or Big Dipper, and also for stars around the constellation of Orion.

The planets
Because the planets are always moving against the background stars, they are treated in some detail in the monthly pages and information is given when they are close to other planets, the Moon, or any of five bright stars that lie near the ecliptic. Such events are known as *appulses* or, more frequently, as *conjunctions*. (There are technical differences in the way these terms are defined – and should be used – in astronomy, but these need not concern us here.)

The term conjunction is also used when a planet is either directly behind or in front of the Sun, as seen from Earth. (Under normal circumstances it will then be invisible.) The conditions of most favourable visibility depend on whether the planet is one of the two known as *inferior planets* (Mercury and Venus) or one of the three *superior planets* (Mars, Jupiter and Saturn) that are covered in detail. Brief details of the fainter superior planets, Uranus and Neptune, are given, especially when they come to opposition.

The inferior planets are most readily seen at eastern or western *elongation*, when their angular distance from the Sun is greatest. For superior planets and minor planets, they are best seen at *opposition*, when they are directly opposite the Sun in the sky, and cross the meridian at local midnight.

Events
A number of interesting events are shown in diagrams for each month. They involve the planets and the Moon, sometimes showing them in relation to specific stars. Events have been chosen as they will appear from one of three different locations: from London; from the central region of the USA; or from Sydney in Australia. Naturally, these events are visible from other locations, but the appearance of the objects on the sky will differ slightly from the diagrams. A list of major astronomical events in 2023 is given on pages 20–21.

Meteors
At some time or other, nearly everyone has seen a *meteor* – a 'shooting star' – as it flashed across the sky. The particles that

*Meteor shower
(showing the April
Lyrid radiant).*

cause meteors – known technically as 'meteoroids' – range in
size from that of a grain of sand (or even smaller) to the size of
a pea. On any night of the year there are occasional meteors,
known as ***sporadics***, that may travel in any direction. These
occur at a rate that is normally between 3 and 8 in an hour. Far
more important, however, are ***meteor showers***, which occur at
fixed periods of the year, when the Earth encounters a trail of
particles left behind by a comet or, very occasionally, by a minor
planet (asteroid). Meteors always appear to diverge from a single
point on the sky, known as the ***radiant***, and the radiants of
major showers are shown on the charts.

Meteors that come from a circular area, 8° in diameter,
around the radiant are classed as belonging to the particular
shower. All others that do not come from that area are sporadics
(or, occasionally, from another shower that is active at the same
time). A list of the major meteor showers is given on the next
page.

Although the positions of the various shower radiants are
shown on the charts, looking directly at the radiant is not the
most effective way of seeing meteors. They are most likely to be
noticed if one is looking about 40–45° away from the radiant
position. (This is approximately two hand-spans as shown in
the diagram for measuring angles on page 13.)

Meteor Showers

Shower	Dates of activity 2023	Date of maximum 2023	Possible hourly rate
Quadrantids	Dec. 28 to Jan. 12	Jan. 3–4	110
α-Centaurids	Jan. 31 to Feb. 20	Feb. 8	6
γ-Normids	Feb. 25 to Mar. 28	Mar. 14–15	6
April Lyrids	Apr. 14–30	Apr. 22–23	18
π-Puppids	Apr. 15–28	Apr. 23–24	var.
η-Aquariids	Apr. 19 to May 28	May 6	50
α-Capricornids	Jul. 3 to Aug. 15	Jul. 30	5
Southern δ-Aquariids	Jul. 12 to Aug. 23	Jul. 30	25
Piscis Austrinids	Jul. 15 to Aug. 10	Jul. 28	5
Perseids	Jul. 17 to Aug. 24	Aug. 12–13	100
α-Aurigids	Aug. 28 to Sep. 5	Sep. 1	6
Southern Taurids	Sep. 10 to Nov. 20	Oct. 10–11	5
Orionids	Oct. 2 to Nov. 7	Oct. 21–22	25
Draconids	Oct. 6–10	Oct. 8–9	10
Northern Taurids	Oct. 20 to Dec. 10	Nov. 12–13	5
Leonids	Nov. 6–30	Nov. 17–18	10
Phoenicids	Nov. 28 to Dec. 9	Dec. 2	var.
Puppid Velids	Dec. 1–15	Dec. 7	10
Geminids	Dec. 4–20	Dec. 14–15	150
Ursids	Dec. 17–26	Dec. 22–23	10

Some Interesting Objects

Messier IC / NGC	Name	Type	Constellation
—	47 Tucanae	globular cluster	Tucana
—	Hyades	open cluster	Taurus
—	Double Cluster	open cluster	Perseus
—	Melotte 111	open cluster	Coma Berenices
M3	—	globular cluster	Canes Venatici
M4	—	globular cluster	Scorpius
M8	Lagoon Nebula	gaseous nebula	Sagittarius
M11	Wild Duck Cluster	open cluster	Scutum
M13	Hercules Cluster	globular cluster	Hercules
M15	—	globular cluster	Pegasus
M22	—	globular cluster	Sagittarius
M27	Dumbbell Nebula	planetary nebula	Vulpecula
M31	Andromeda Galaxy	galaxy	Andromeda
M35	—	open cluster	Gemini
M42	Orion Nebula	gaseous nebula	Orion
M44	Praesepe	open cluster	Cancer
M45	Pleiades	open cluster	Taurus
M57	Ring Nebula	planetary nebula	Lyra
M67	—	open cluster	Cancer
IC 2602	Southern Pleiades	open cluster	Carina
NGC 752	—	open cluster	Andromeda
NGC 3242	Ghost of Jupiter	planetary nebula	Hydra
NGC 3372	Eta Carinae Nebula	gaseous nebula	Carina
NGC 4755	Jewel Box	open cluster	Crux
NGC 5139	Omega Centauri	globular cluster	Centaurus

Other objects

Certain other objects may be seen with the naked eye under good conditions. Some were given names in antiquity – Praesepe is one example – but many are known by what are called 'Messier numbers', the numbers in a catalogue of nebulous objects compiled by Charles Messier in the late-eighteenth century. Some, such as the Andromeda Galaxy, M31, and the Orion Nebula, M42, may be seen faintly by the naked eye, but all those given in the list here will benefit from the use of binoculars.

Apart from galaxies, such as M31, which contain thousands of millions of stars, there are also two types of cluster: open clusters, such as M45, the Pleiades, which may consist of a few dozen to some hundreds of stars; and globular clusters, such as M13 in Hercules, which are spherical concentrations of many thousands of stars. One or two gaseous nebulae, consisting of gas illuminated by stars within them are also visible. The Orion Nebula, M42, is one, and is illuminated by the group of four stars, known as the Trapezium, which may be seen within it by using a good pair of binoculars. A list of interesting objects is given on the previous page.

In 1781, **Charles Messier** (26 June 1730 – 12 April 1817) published the final version of his catalogue of 110 nebulous objects and faint star clusters that might be confused with comets. The objects in this catalogue are still know as the Messier objects and are always quoted as 'M' numbers. Some of the most famous are M1, the Crab Nebula; M31, the Andromeda galaxy; M42, the Orion Nebula; and M45, the Pleiades.

Dates and time

Astronomers, worldwide, use a standardized method of expressing the date and time. This prevents confusion in comparing observations made by different observers. The various elements are given in descending order: year, month (three-letter abbreviation to prevent confusion over using numbers), day, hour, minutes, seconds. (In extreme cases, fractions of minutes or seconds may be used.) The date and time are that on the Greenwich meridian (GMT), and ignore any changes for Summer Time / Daylight Saving Time (DST), and any adjustments for local time at the observer's location. This standard is known as Coordinated Universal Time (UTC). In this book and in many others this is generally given as Universal Time (UT). All times given in this book are in UT.

To avoid problems over the changes involved in moving to and from Summer Time / Daylight Saving Time (and the complications over the beginning and end dates) and also the adjustments for local time, experienced astronomers set a (cheap) watch or clock to Universal Time and keep it that way. Smartphone users may use a simple world clock app, provided they lock it to the time (GMT) on the Greenwich meridian.

Similarly, the date given for an event is the date as it applies at the Greenwich meridian, i.e., in UT. Occasionally, this may differ from the date as given by your local time. An event that occurs (say) late in the night in Europe may seem to occur on the previous day to an observer to the west (such as in the USA), when local time is taken into account. This is another complication that is avoided by using the Universal Time standard.

By 1845, five asteroids had been discovered; by 1868, 100 were known. Numbers increased dramatically following the introduction of photographic methods and dedicated searches and many thousands are now known. (1000 by 1920; 100,000 by 2005 and over one million to date.) The majority of these objects orbit in the inner Solar System, in the Main Belt between Mars and Jupiter, with the Trojan asteroids locked into Jupiter's orbit. There are additional 'families' of asteroids whose orbits take them past the inner planets, Earth, Venus and Mercury. There is also a large population in the outer Solar System, beyond the orbit of Neptune, these objects being known as Trans-Neptunian Objects (TNOs). There are various names for this outer belt of objects, part being known as the Kuiper Belt.

Major Events in 2023

Jan. 03	Quadrantid meteor shower maximum
Jan. 04	Earth at perihelion (closest to the Sun)
Jan. 08	Minor planet (2) Pallas at opposition
Jan 26	Minor planet (6) Hebe at opposition
Jan. 30	Mercury at greatest elongation west
Feb. 08	α-Centaurid meteor shower maximum
Mar. 15	γ-Normid meteor shower maximum
Mar. 20	Northern spring / Southern autumnal equinox
Mar. 21	Dwarf planet Ceres at opposition
Apr. 11	Mercury at greatest elongation east
Apr. 20	Hybrid solar eclipse over Indonesia
Apr. 22	April Lyrid meteor shower maximum
Apr. 24	π-Puppid meteor shower maximum
May 05	Penumbral lunar eclipse
May 06	Eta Aquariid meteor shower maximum
May 29	Mercury at greatest elongation west
Jun. 04	Venus at greatest elongation east
Jul. 06	Earth at aphelion (farthest from the Sun)
Jul. 07	Minor planet (15) Eunomia at opposition
Jul. 28	Piscis Austrinid meteor shower maximum
Jul. 30	α-Capricornid meteor shower maximum
Jul. 30	Southern δ-Aquariid meteor shower maximum
Aug. 10	Mercury at greatest elongation east

Major events in 2023 (continued)

Aug. 12	Perseid meteor shower maximum
Aug. 27	Minor planet (8) Flora at opposition
Aug. 27	Saturn at opposition
Sep. 01	α-Aurigid meteor shower maximum
Sep. 19	Neptune at opposition
Sep. 22	Mercury at greatest elongation west
Sep. 23	Northern autumnal / southern spring equinox
Oct. 08	Draconid meteor shower maximum
Oct. 10	Southern Taurid meteor shower maximum
Oct. 14	Annular solar eclipse
Oct. 21	Orionid meteor shower maximum
Oct. 23	Venus at greatest elongation west
Oct. 28	Partial lunar eclipse
Nov. 03	Jupiter at opposition
Nov. 12	Northern Taurid meteor shower maximum
Nov. 13	Uranus at opposition
Nov. 17	Leonid meteor shower maximum
Dec. 04	Mercury at greatest elongation east
Dec. 14	Geminid meteor shower maximum
Dec. 21	Minor planet (4) Vesta at opposition
Dec. 22	Ursid meteor shower maximum
Dec. 22	Minor planet (9) Metis at opposition

The Moon

The Moon at First Quarter.

The Moon

The monthly pages include diagrams showing the **phase** of the Moon (see page 28) for every day of the month, and also indicate the day in the **lunation** (or **age** of the Moon), which begins at New Moon. The diagrams showing the Moon's phase are repeated for southern-hemisphere observers who will see the Moon, south up. Although the main features of the surface – the light highlands and the dark maria (seas) – may be seen with the naked eye, far more features may be detected with the use of binoculars or any telescope. The many craters are best seen when they are close to the **terminator** (the boundary between the illuminated and the non-illuminated areas of the surface), when the Sun rises or sets over any particular region of the Moon and the crater walls or central peaks cast strong shadows. Most features become difficult to see at Full Moon, although this is the best time to see the bright ray systems surrounding certain craters. Accompanying the Moon map on the following pages is a list of prominent features, including the days in the lunation when features are normally close to the terminator and thus easiest to see. A few bright features such as Linné and Proclus, visible when well illuminated, are also listed. One feature, Rupes Recta (the Straight Wall) is readily visible only when it casts a shadow with light from the east, appearing as a light line when illuminated from the opposite direction.

The dates of visibility vary slightly through the effects of **libration**. Because the Moon's orbit is inclined to the Earth's equator and also because it moves in an ellipse, the Moon appears to rock slightly from side to side (and nod up and down). Features near the **limb** (the edge of the Moon) may vary considerably in their location and visibility. (This is easily noticeable with Mare Crisium and the craters Tycho and Plato.) Another effect is that at crescent phases before and after New Moon, the normally non-illuminated portion of the Moon receives a certain amount of light, reflected from the Earth. This **Earthshine** may enable certain bright features (such as Aristarchus, Kepler and Copernicus) to be detected even though they are not illuminated by sunlight.

Moon features

The numbers below indicate the age of the Moon when features are usually best visible.

Abulfeda	6:20	Gassendi	11:25	Philolaus	9:23
Agrippa	7:21	Geminus	3:17	Piccolomini	5:19
Albategnius	7:21	Goclenius	4:18	Pitatus	8:22
Aliacensis	7:21	Grimaldi	13–14:27–28	Pitiscus	5:19
Alphonsus	8:22	Gutenberg	5:19	Plato	8:22
Anaxagoras	9:23	Hercules	5:19	Plinius	6:20
Anaximenes	11:25	Herodotus	11:25	Posidonius	5:19
Archimedes	8:22	Hipparchus	7:21	Proclus	14:18
Aristarchus	11:25	Hommel	5:19	Ptolemaeus	8:22
Aristillus	7:21	Humboldt	3:15	Purbach	8:22
Aristoteles	6:20	Janssen	4:18	Pythagoras	12:26
Arzachel	8:22	Julius Caesar	6:20	Rabbi Levi	6:20
Atlas	4:18	Kepler	10:24	Reinhold	9:23
Autolycus	7:21	Landsberg	10:24	Rima Ariadaeus	6:20
Barrow	7:21	Langrenus	3:17	Rupes Recta	8
Billy	12:26	Letronne	11:25	Saussure	8:22
Birt	8:22	Linné	6	Scheiner	10:24
Blancanus	9:23	Longomontanus	9:23	Schickard	12:26
Bullialdus	9:23	Macrobius	4:18	Sinus Iridum	10:24
Bürg	5:19	Mädler	5:19	Snellius	3:17
Campanus	10:24	Maginus	8:22	Stöfler	7:21
Cassini	7:21	Manilius	7:21	Taruntius	4:18
Catharina	6:20	Mare Crisium	2–3:16–17	Thebit	8:22
Clavius	9:23	Maurolycus	6:20	Theophilus	5:19
Cleomedes	3:17	Mercator	10:24	Timocharis	8:22
Copernicus	9:23	Metius	4:18	Triesnecker	6–7:21
Cyrillus	6:20	Meton	6:20	Tycho	8:22
Delambre	6:20	Mons Pico	8:22	Vallis Alpes	7:21
Deslandres	8:22	Mons Piton	8:22	Vallis Schröteri	11:25
Endymion	3:17	Mons Rümker	12:26	Vlacq	5:19
Eratosthenes	8:22	Montes Alpes	6–8:21	Walther	7:21
Eudoxus	6:20	Montes Apenninus	8	Wargentin	12:27
Fra Mauro	9:23	Orontius	8:22	Werner	7:21
Fracastorius	5:19	Pallas	8:22	Wilhelm	9:23
Franklin	4:18	Petavius	3:17	Zagut	6:20

Map of the Moon

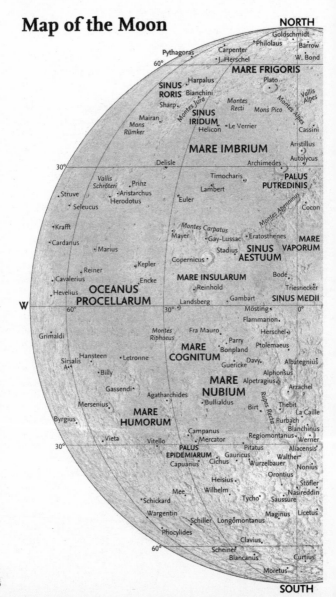

NORTH

Goldschmidt
Philolaus · Barrow
Pythagoras · Carpenter · W. Bond
· J. Herschel

60°

MARE FRIGORIS

Harpalus · Plato
SINUS
RORIS Bianchini
Sharp · *Montes* · Vallis
Montes Jura *Recti* Mons Pico *Alpes*
Mairan **SINUS** · Le Verrier *Montes Alpes*
Mons **IRIDUM** Helicon Cassini
Rümker

MARE IMBRIUM
Aristillus
30° Delisle Archimedes Autolycus

Vallis Timocharis **PALUS**
Schröteri Prinz Lambert **PUTREDINIS**
· Struve · Aristarchus
· Seleucus Herodotus Euler Cocon

· Krafft *Montes Apenninus*
· Cardanus *Montes Carpatus* **MARE**
Mayer · Gay-Lussac · Eratosthenes **VAPORUM**
· Marius Stadius **SINUS**
Reiner Copernicus · **AESTUUM** Bode·
Kepler Triesnecker·
· Cavalerius Encke **MARE INSULARUM** **SINUS MEDII**
Hevelius **OCEANUS** · Reinhold
PROCELLARUM Landsberg · Gambart Mösting·
W 60° 30° 0°
Flammarion·
Grimaldi *Montes* Fra Mauro Herschel·
Riphaeus · Parry Ptolemaeus
· Hansteen · Letronne **MARE** Bonpland
Sirsalis · **COGNITUM** Guericke Davy· Albategnius·
A · Billy Alphonsus
Gassendi· **MARE** Alpetragius· Arzachel
Mersenius · Agatharchides **NUBIUM** Thebit
· Bullialdus Birt· *Rupes Recta* La Caille
Byrgius · **MARE** Purbach
HUMORUM Campanus Regiomontanus· Blanchinus
· Vieta Vitello Mercator **PALUS** Pitatus Werner
30° **EPIDEMIARUM** Gauricus Walther Aliacensis·
Capuanus· Cichus Wurzelbauer Nonius
Orontius Stöfler·
Heisius · Nasireddin
Mee Wilhelm Saussure
· Schickard Tycho· Maginus Licetus
Wargentin Longomontanus
Schiller Clavius·
Phocylides
60°
Scheiner
Blancanus Curtius·
Moretus·

SOUTH

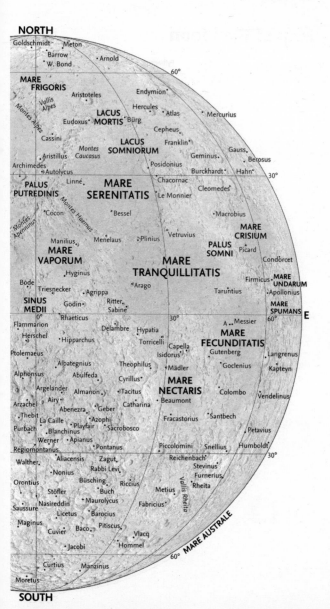

NORTH

Goldschmidt · Meton
Barrow
W. Bond · Arnold

MARE
FRIGORIS

Aristoteles
Vallis
Alpes
Eudoxus
Cassini

LACUS
MORTIS

Endymion
Hercules
Atlas
Cepheus
Franklin
Geminus
Burckhardt
Mercurius

Gauss
Berosus
Hahn

60°

Montes Alpes

LACUS
SOMNIORUM

Archimedes
Aristillus
Autolycus

Montes
Caucasus

Linné

Posidonius
Chacornac
Le Monnier

Macrobius

30°

PALUS
PUTREDINIS

MARE
SERENITATIS

Cleomedes

Montes Haemus

Cocon
Bessel

MARE
CRISIUM

Montes
Apenninus

Manilius
Menelaus
Plinius
Vetruvius

PALUS
SOMNI

Picard

Condorcet

MARE
VAPORUM

Hyginus

MARE
TRANQUILLITATIS

Firmicus

MARE
UNDARUM

Bode
Triesnecker
Agrippa
Arago
Taruntius
Apollonius

SINUS
MEDII

Godin
Ritter
Sabine

MARE
SPUMANS

0°
Rhaeticus

30°

A · Messier

60°

E

Flammarion
Herschel
Hipparchus

Delambre
Hypatia
Torricelli
Capella
Isidorus

MARE
FECUNDITATIS

Langrenus

Ptolemaeus

Albategnius
Theophilus
Mädler

Gutenberg
Goclenius

Kapteyn

Alphonsus
Abulfeda
Cyrillus

MARE
NECTARIS

Argelander
Almanon
Tacitus

Arzachel
Airy
Geber
Catharina

Abenezra
Geber

Beaumont
Colombo

Vendelinus

Thebit
La Caille
Azophi
Playfair
Sacrobosco

Fracastorius
Santbech

Purbach
Blanchinus

Petavius

Werner
Apianus
Pontanus
Piccolomini
Snellius
Humboldt

Regiomontanus

30°

Walther
Aliacensis
Zagut
Reichenbach

Nonius
Rabbi Levi
Stevinus

Orontius
Büsching
Riccius
Furnerius
Rheita

Stöfler
Buch
Metius

Nasireddin
Maurolycus
Fabricius

Saussure
Licetus
Barocius

Maginus
Cuvier
Baco
Pitiscus

Jacobi
Vlacq
Hommel

60°

MARE AUSTRALE

Curtius
Manzinus

Moretus

SOUTH

Moon phases

The diagram on this page shows how the different appearance of the Moon occurs. The changes in its apparent shape are purely related to the position of the Moon in its orbit around the Earth. They are not, as some people mistakenly believe, caused by the shadow of the Earth on the Moon. (The only time that happens is during a rare lunar eclipse, as described on pages 10–11.) Such events occur only at Full Moon, when the Sun and Moon are on opposite sides of the Earth. The diagram shows that New Moon is when the Moon lies between the Sun and Earth, and Full Moon when the Earth is between the Sun and the Moon. The age of the Moon (in days) is reckoned from the time of New Moon. After New Moon we have a waxing crescent until First Quarter, after which the Moon is described as waxing gibbous. After Full Moon we have a waning gibbous Moon until Last Quarter. Following that, we have a waning crescent until the next New Moon, when the sequence repeats.

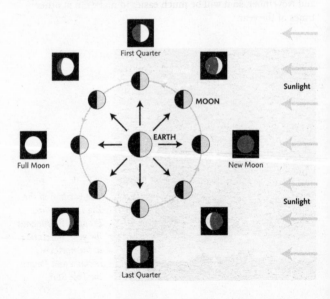

The Circumpolar Constellations

The northern circumpolar constellations

Learning the patterns of the stars, the constellations and asterisms is not particularly difficult. You need to start by identifying the various constellations that are circumpolar where you live. These are always above the horizon, so you can generally start at any time of the year. The charts on pages 30 and 32 show the northern and southern circumpolar constellations, respectively. The fine, dashed lines indicate the areas that are circumpolar at different latitudes.

The key constellation when learning the pattern of stars in the northern sky is **Ursa Major**, in particular the seven stars forming the asterism known to many as the **'Plough'** or to people in North America as the **'Big Dipper'**. As the chart shows, this is just circumpolar for anyone at latitude 40°N, except for **Alkaid** (η Ursae Majoris), the last star in the 'tail'. Even so, the asterism of the Plough is low on the northern horizon between September and November, so it will be much easier to make out at other times of the year.

The position of the Big Dipper (the Plough), throughout the year, in relation to the northern horizon and Polaris, the Pole Star.

The northern circumpolar constellations.

The two stars **Dubhe** and **Merak** (α and β Ursae Majoris) are known as the 'Pointers', because they indicate the position of **Polaris**, the Pole Star (α Ursae Minoris), at about a distance of five times their separation. Following this line takes you to the constellation of **Ursa Minor**, the 'Little Bear' or 'Little Dipper', where Polaris is at the end of the 'tail' or 'handle'.

On the far side of the Pole is the constellation of **Cassiopeia**, which is highly distinctive, with its five main stars forming the letter 'M' or 'W', depending on its orientation. Cassiopeia is circumpolar for observers at latitude 40°N or closer to the North Pole, although at times it is near the northern horizon and more difficult to see. (But at such times Ursa Major is clearly visible.) To find Cassiopeia from Ursa Major, start at **Alioth** (ε Ursae Majoris) and extend a line from that star to Polaris and beyond. It points to the central star of the five.

Moving anticlockwise from Cassiopeia, we come to **Cepheus**, which has been likened to the gable end of a house, with its base in the Milky Way. The line from the Pointers to Polaris, if extended points to **Errai** (γ Cephei), at the top of the 'gable'. Continuing in the same direction, we come to **Draco,** which wraps around Ursa Minor. The quadrilateral of stars that forms the 'head' of Draco is just circumpolar for observers at 40°N latitude, although it is brushing the horizon in January. On the opposite side of the sky to the head of Draco, the whole of the faint constellation of **Camelopardalis** is visible.

For observers slightly farther north, say at 50°N, additional constellations become circumpolar. The most important of these are **Perseus**, not far from Cassiopeia, and most of which is visible and, farther round, the northern portion of **Auriga**, with bright **Capella** (α Aurigae). On the other side of the sky is **Deneb**, the brightest star in **Cygnus**, although it is often close to the horizon, especially during the early night during the winter months. **Vega** (α Lyrae) another of the three stars that form the Summer Triangle is even farther south, often brushing the northern horizon, and only truly circumpolar and clearly seen at any time of the year for observers at 60°N.

Such far northern observers will also find that **Castor** (α Geminorum) is actually circumpolar, although at times it is extremely low on the horizon. The other bright star in **Gemini**, **Pollux** (β Geminorum) is slightly farther south and cannot really be considered circumpolar.

The southern circumpolar constellations.

Eta Carinae (η Carinae) is one of the most massive and luminous stars known. It is estimated to have a mass between 120 and 150 times that of the Sun, and be between four and five million times as luminous.

The southern circumpolar constellations

Just as Ursa Major is the key constellation in the northern sky,
so is **Crux** (the Southern Cross) an easily recognized feature
of the southern circumpolar sky, although at times it may be
brushing the horizon for observers at 30°S – roughly the latitude
of Sydney in Australia. This is particularly true in the southern
spring. Northeners, new to the southern sky, sometimes mistake
the '**False Cross**', which consists of two stars from each of the
constellations of **Vela** and **Carina** for the true Southern Cross.
Crux itself is accompanied by the clearly visible dark cloud of
the 'Coalsack' and also the two brightest stars in **Centaurus: Rigil
Kentaurus** and **Hadar** (α and β Centauri, respectively). Together,
the four stars of Crux and the two from Centaurus act as
principal guides to the southern constellations.

Unfortunately, unlike the situation in the north, there is no
star conveniently located at the South Celestial Pole (SCP), which
lies in a relatively empty region of sky in the faint, triangular
constellation of **Octans**. Octans itself is perhaps best found by
using the stars of **Pavo** as guides. A line from **Peacock** (α Pavonis)
through the second brightest star (β) in that constellation, if

*The position of
Crux, the Southern
Cross, throughout
the year, in relation
to the southern
horizon. It also
shows the position
of the two brightest
stars in Centaurus.*

33

extended by about the same amount as the distance between the two stars, points to the 'base' of the triangle of Octans.

The main 'upright' of Crux, if extended and curving slightly to the right, does point in the approximate direction of the Pole, passing through *Musca* and the tip of *Chamaeleon*. However, a better way is to start at Hadar (the star in the bright pair that is closest to Crux), turn at right-angles at Rigil Kentaurus, and following an imaginary line through the brightest star in the small constellation of *Circinus* and then right across the sky, brushing past the outlying star of *Apus*, and the star (δ) at the apex of Octans itself.

Centaurus is a straggling constellation, with many stars well north of Rigil Kentaurus and Hadar, and some fainter ones that partially enclose Crux. Starting at Crux, and moving clockwise (in the same direction as the sky rotates), we come to the stars of *Carina* and *Vela*, both originally part of the large, now obsolete constellation of Argo Navis. Past the False Cross, we come to *Canopus* (α Carinae), the second brightest star in the sky, which is just circumpolar for observers at 40°S, although occasionally, especially in June, very low over the northern horizon. Lying between Canopus and the SCP is the **Large Magellanic Cloud** (LMC), a satellite galaxy to our own. It lies across the boundaries of the constellations of *Dorado* and *Mensa*.

Continuing round from Canopus we pass the constellation of Dorado, the small constellation of *Reticulum* and the undistinguished constellation of **Horologium**, beyond which is **Achernar** (α Eridani) the brightest star in the long, winding constellation of **Eridanus**, which actually starts far to the north, close to **Rigel** in **Orion**. Between Achernar and the SCP lies the triangular constellation of **Hydrus**, next to the constellation of **Tucana** which contains the **Small Magellanic Cloud** (SMC).

For observers farther south (at say, 50°S) there are the constellations of **Phoenix**, followed by the roughly cross-shaped constellation of **Grus** and the rather faint **Indus**. For observers at 40°S, the whole of the constellation of Pavo is visible, including its brightest star, Peacock. Farther round there is the constellation of **Ara** and, for observers farther north at 30°S, **Triangulum Australe** is fully visible.

Of the 88 constellations, the largest is **Hydra** with an area of 1303 square degrees. It covers more than 7 hours of right ascension (105° in the sky). It is more than 19 times as large as the smallest, **Crux**, which has an area of just 68 square degrees. **Equuleus** (87th), is not much larger with an area of 72 square degrees. **Sagitta** is the 86th, area 80 square degrees, and **Circinus**, 85th, with an area of 93 square degrees. None of these four small constellation includes any objects of particular interest.

The Monthly Maps

How to use the monthly maps

The charts in this book are designed to be used more-or-less anywhere in the world. They are not suitable to be used at very high northern or southern latitudes (beyond 60°N or 60°S). That is slightly less than the latitudes of the Arctic and Antarctic Circles, beyond which there are approximately six months of daylight, followed by six months of darkness. The design may seem a little complicated, but these diagrams should make their usage clear. The main charts are given in pairs, one pair for each month: Looking North and Looking South.

Obviously, the region of the sky that is visible at any time entirely depends on one's location on Earth. You should imagine a rectangular 'window', 90° degrees high, that includes the sky from the horizon to the zenith. Think of moving this 'window' north or south over the charts, depending on your actual latitude. The base will be at your actual latitude on Earth. The other edge will be at your zenith. (You could make an actual 'window', by drawing a rectangle on a thin sheet of plastic, with the horizon and zenith lines 90° apart, and then use this on the charts.) The diagram on the next page shows this 'window' for the latitude of 50°N.

The scales on the right-hand and left-hand margins indicate the northern (or southern) horizon, for looking north (or south), respectively. The two diagrams on page 38 are drawn to indicate the horizon for the latitude of 40°N (the latitude of Philadelphia in the United States or Madrid in Spain); the second pair on page 39 show what would be visible 'looking north' and 'looking south' from latitude 30°S (the latitude of Durban in South Africa). If you are looking north (or south), once you get to the zenith, you can switch to the other chart, showing the view from the southern (or northern) horizon to the zenith.

To help you choose the correct latitude, there is a world map on pages 40–41.

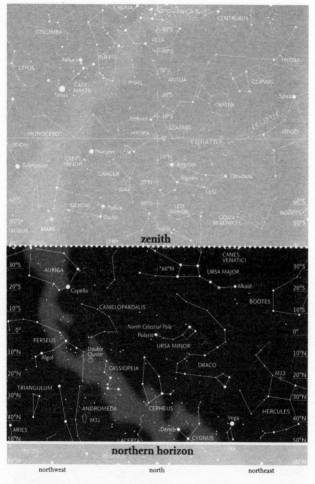

Horizon window, from the northern horizon (solid line at the bottom) to the zenith (the dotted line) for the latitude of 50°N.

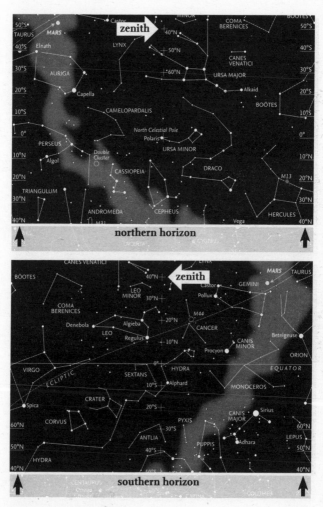

Horizons for latitude 40°N.

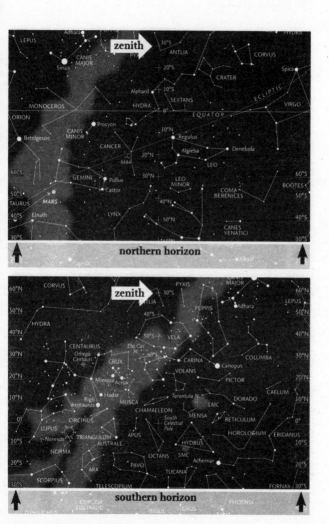

Horizons for latitude 30°S.

World Map

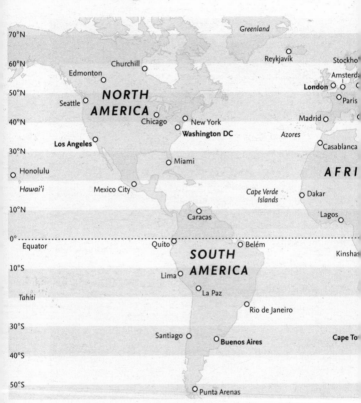

January

January

January – Introduction

The Earth

The Earth reaches perihelion (the closest point to the Sun in its yearly orbit) on January 4 at 16:17 Universal Time, when its distance from the Sun is 0.983295578 AU (147,098,855 km).

The Moon occults Mars on January 3, but this event is only visible from the southern Indian Ocean and southern Africa (see page 51). Mars will then be in Taurus. Another occultation of Mars will occur on January 31. This will be visible from a wide region of Central America and the southwestern United States. For northern-hemisphere observers the *Quadrantid* meteor shower reaches maximum on the night of January 3–4, when the Moon is waxing gibbous, just 2–3 days before Full Moon, so conditions are not very favourable. It is one of the strongest showers of the year, with rates comparable with the Perseids (in August) and the Geminids (in December). However, peak activity is very short and easily missed, although the stream occasionally produces bright fireballs.

The planets

Mercury passes inferior conjunction on January 7. Although it is bright (mag. -3.9), *Venus* is close to the Sun in twilight. *Mars* is in Taurus, slowly fading from mag. -1.2 to -0.3 over the month. It is occulted by the Moon on January 3 and 31. *Jupiter* is in *Pisces* (average mag. -2.4) and *Saturn* (mag. 0.8) in *Capricornus*, close to the Sun in evening twilight. *Uranus*, initially retrograding until January 23, is mag. 5.7 in *Aries*. *Neptune* is mag. 7.9 in *Aquarius*. On January 8, the minor planet (2) *Pallas* is at opposition in *Canis Major* at mag. 7.7 (see the charts on the facing page). On January 26, minor planet (6) *Hebe* is at opposition in *Hydra* at mag. 8.8.

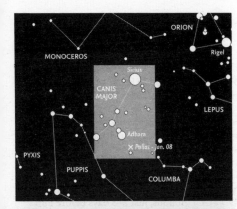

A finder chart for the position of minor planet (2) Pallas, at its opposition. The grey area is shown in more detail on the map below.

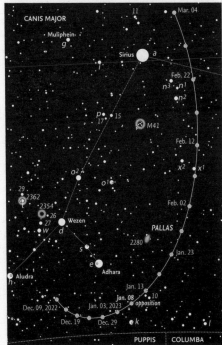

The path of the minor planet (2) Pallas around its opposition on January 8 (mag. 7.7). Background stars are shown down to magnitude 8.0.

45

Sunrise and sunset

City	Date	Sunrise	Sunset
Buenos Aires, Argentina			
	Jan. 01	08:44	23:10
	Jan. 31	09:12	23:01
Cape Town, South Africa			
	Jan. 01	03:38	18:01
	Jan. 31	04:06	17:52
London, UK			
	Jan. 01	08:07	16:02
	Jan. 31	07:42	16:48
Los Angeles, USA			
	Jan. 01	14:59	00:54
	Jan. 31	14:51	01:21
Nairobi, Kenya			
	Jan. 01	03:30	15:42
	Jan. 31	03:41	15:51
Sydney, Australia			
	Jan. 01	18:48	09:09
	Jan. 31	19:16	09:01
Tokyo, Japan			
	Jan. 01	21:51	07:38
	Jan. 31	21:41	08:07
Washington, DC, USA			
	Jan. 01	12:27	21:57
	Jan. 31	12:15	22:29
Wellington, New Zealand			
	Jan. 01	16:52	07:57
	Jan. 31	17:26	07:43

NB: *the times given are in Universal Time (UT)*

The Moon's phases and ages

Northern hemisphere

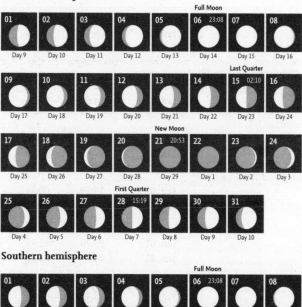

Southern hemisphere

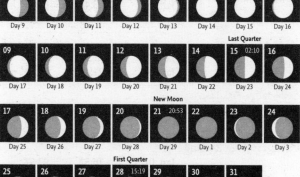

The Moon

The Moon in January

On January 1, the Moon passes 0.7° north of **Uranus** (mag. 5.7). On January 3 it occults **Mars** in **Taurus** (visible from the Indian Ocean and southern Africa), and the next day is 8.1° north of **Aldebaran**. On January 7, it is 1.9° south of **Pollux** in **Gemini**. On January 10, it passes 4.6° north of **Regulus**, between it and **Algieba**. By January 14, a day before Last Quarter it will be 3.8° north of **Spica** in **Virgo**. As a waning crescent it will be 2.1° north of **Antares** on January 18. On January 20, one day before New Moon, it is 6.9° south of **Mercury** (mag. -0.6) just past inferior conjunction and invisible in the evening sky. On January 23, the thin waxing crescent will be 3.8° south of **Saturn** (mag. 0.8), and a little later, 3.5° south of **Venus**, much brighter at mag. -3.9. On January 31 it occults **Mars** again, this time visible from Central America and the southwestern United States.

Wolf Moon

Because the howling of wolves is often heard in North America in winter, the Full Moon in January is often known as the 'Wolf Moon'. The name may originally stem from the Old-World, Anglo-Saxon lunar calendar. Other names for this Full Moon include: Moon After Yule, Old Moon, Ice Moon, and Snow Moon. Among the Algonquin tribes the name was 'squochee kesos', meaning 'the Sun has not strength to thaw'. The name 'Wolf Moon' was occasionally applied to the Full Moon in December.

In this photograph, the narrow lunar crescent (about two days old) has been over-exposed to show the Earthshine illuminating on the other portion of the Moon, where the dark maria are faintly visible.

Names of the Full Moon in North America
Many different cultures had specific names for the Full Moon, depending on the time of year. Even in a single culture, the actual names often varied between different tribes, so there may be more than one name used for a particular Full Moon. The interval between successive Full Moons (or between any other specific phases of the Moon) is known as the synodic month, and is, on average, 29.53 days, so the names have come to be associated with modern calendar months. One of the most commonly known sets of names is that used by the various tribes in North America. These are:

- January: 'Wolf Moon'
- February: 'Snow Moon', 'Hunger Moon', 'Storm Moon'
- March: 'Worm Moon', 'Crow Moon', 'Sap Moon', 'Lenten Moon'
- April: 'Seed Moon', 'Pink Moon', 'Sprouting Grass Moon', 'Egg Moon', 'Fish Moon'
- May: 'Milk Moon', 'Flower Moon', 'Corn Planting Moon'
- June: 'Mead Moon', 'Strawberry Moon', 'Rose Moon', 'Thunder Moon'
- July: 'Hay Moon', 'Buck Moon', 'Elk Moon', 'Thunder Moon'
- August: 'Corn Moon', 'Sturgeon Moon', 'Red Moon' 'Green Corn Moon', 'Grain Moon'
- September: 'Harvest Moon', 'Full Corn Moon'
- October: 'Hunter's Moon', 'Blood Moon'/'Sanguine Moon'
- November: 'Beaver Moon', 'Frosty Moon'
- December: 'Oak Moon', 'Cold Moon', 'Long Night's Moon'

The only two names commonly used in Europe were 'Harvest Moon' and 'Hunter's Moon'. On rare occasions, particularly in religious contexts, the term 'Lenten Moon' was used for the Full Moon in March. The other terms, which originated in North America, have been adopted increasingly by the media in recent years.

Calendar for January

01–12		Quadrantid meteor shower
01	22:16	Uranus 0.7°S of the Moon
03	19:38	Mars occulted by the Moon
03–04		Quadrantid meteor shower maximum
04	00:53	Aldebaran 8.1°S of the Moon
04	16:17	Earth at perihelion (0.983295578 AU)
06	23:08	Full Moon
07	12:57	Mercury at inferior conjunction
07	14:18	Pollux 1.9°N of the Moon
08	09:19	Moon at apogee = 406,458 km
08	19:02	Minor planet (2) Pallas at opposition (mag. 7.7)
10	12:17	Regulus 4.6°S of the Moon
14	22:37	Spica 3.8°S of the Moon
15	02:10	Last Quarter
18	10:04	Antares 2.1°S of the Moon
20	07:49	Mercury 6.9°N of the Moon
21	20:53	New Moon
21	20:57	Moon at perigee = 356,569 km
22	20:00 *	Saturn (mag. 0.8) 0.4°N of Venus (mag. -3.9)
23	07:21	Saturn 3.8°N of the Moon
23	08:18	Venus 3.5°N of the Moon
25	05:55	Neptune 2.7°N of the Moon
26	02:03	Jupiter 1.8°N of the Moon
26	09:17	Minor planet (6) Hebe at opposition (mag. 8.8)
28	15:19	First Quarter
29	04:09	Uranus 1.0°S of the Moon
30	05:54	Mercury at greatest elongation (25.0°W, mag. -0.1)
31–Feb.20		α-Centaurid meteor shower
31	04:25	Mars occulted by the Moon
31	06:45	Aldebaran 8.3°S of the Moon

These objects are close together for an extended period around this time.

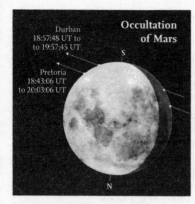

January 3 · *Mars is occulted by the Moon. Times of disappearance and reappearance are given for Durban and Pretoria (as seen from South Africa).*

January 7 · *The Moon rises in the company of Castor and Pollux (as seen from London).*

January 22–23 · *The crescent Moon passes Venus and Saturn (as seen from central USA).*

January 31 · *Mars is occulted by the Moon. Times of disappearance and reappearance are given for Houston and Mexico City (as seen from central USA).*

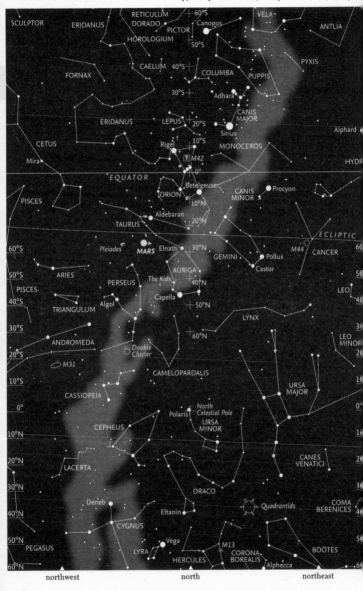

northwest   north   northeast

January – Looking North

For most northern observers, all the important northern circumpolar constellations (see pages 29–31) of **Ursa Major**, **Ursa Minor** (with **Polaris**, the Pole Star), **Cassiopeia**, **Draco** and **Cepheus** will be visible. Polaris is, of course, the star about which the sky appears to rotate, even though it is not precisely at the North Celestial Pole. Ursa Major stands more-or-less vertically above the horizon in the northeast. Opposite it in the northwest of the sky is the 'W' of Cassiopeia. (Looking more like an 'M' at this time of the year.) For most observers, **Capella** (α Aurigae) is high overhead, but only those at high latitudes will find it easy to see the quadrilateral of stars that marks the head of Draco, brilliant **Deneb** (α Cygni) or the even brighter **Vega** (α Lyrae), yet farther south. Deneb and **Eltanin** (γ Draconis), the brightest star in the 'Head' of Draco, are skimming the horizon for observers at 40°N.

On 2 January 2019, the Chinese lunar probe **Chang-e 4** became the first object to achieve a landing on the far side of the Moon. It landed near the South Pole in the Von Karman crater, which is itself in the South Pole-Aitken basin.

For observers between about 30 and 50°N, the constellation of **Auriga** is near the zenith (and thus difficult to observe). This important constellation contains bright **Capella** (α Aurigae) and farther to its west lies the constellation of **Perseus**, with **Algol**, the famous variable star.

Capella

• The Kids

AURIGA

Elnath (β Tau) •

The constellation of Auriga, with brilliant Capella which, although appearing as a single star, is actually a quadruple system, consisting of a pair of yellow giant stars, gravitationally bound to a more distant pair of red dwarfs. Elnath, near the bottom, actually belongs to the constellation of Taurus.

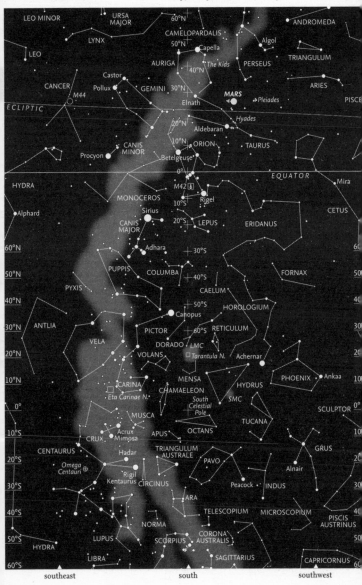

January – Looking South

The southern sky is dominated by *Orion*, visible from nearly everywhere in the world and prominent during the northern winter months. For observers near the equator it is, of course, high above near the zenith. Orion is highly distinctive, with a line of three stars that form the 'Belt'. To most observers, the bright star *Betelgeuse* (α Orionis), shows a reddish tinge, in contrast to the brilliant bluish-white *Rigel* (β Orionis). The three stars of the belt lie directly south of the celestial equator. A vertical line of three 'stars' forms the 'Sword' that hangs south of the Belt. With good viewing, the central 'star' appears as a hazy spot, even to the naked eye, and is actually the *Orion Nebula* (M42). Binoculars reveal the four stars of the Trapezium, which illuminate the nebula.

Orion's Belt points up to the northwest towards *Taurus* (the Bull) and orange-tinted *Aldebaran* (α Tauri). Close to Aldebaran, there is a conspicuous 'V' of stars, called the *Hyades* cluster. (Despite appearances, Aldebaran is not part of the cluster.) Farther along, the same line from Orion passes below a bright cluster of stars, the *Pleiades*, or Seven Sisters. Even the smallest pair of binoculars reveals this as a beautiful group of bluish-white stars. The two most conspicuous of the other stars in Taurus lie directly north of Orion, and form an elongated triangle with Aldebaran. The northernmost, *Elnath* (β Tauri), was once considered to be part of the constellation of *Auriga*.

Slightly to the west of *Capella* lies a small triangle of fainter stars, known as '*The Kids*'. (Ancient mythological representations of Auriga show him carrying two young goats.) Together with Elnath, the body of Auriga forms a large pentagon on the sky, with The Kids lying on the western side (see page 53).

Running south from Orion is the long constellation of *Eridanus* (the River), which begins near Rigel in Orion and runs far south to end at *Achernar* (α Eridani). To the south of Orion is the constellation of *Canis Major* and several other constellations, including the oddly shaped *Carina*. The line of Orion's Belt also points southeast in the general direction of *Sirius* (α Canis Majoris), the brightest star. Almost due south of Sirius lies *Canopus* (α Carinae), the second brightest star in the sky.

Motion of the planets

Both the planets Mars and Uranus end their retrograde motion in January 2023 and revert to direct motion.

Retrograde motion of the planets was one of the main problems that faced ancient astronomers and astrologers when it was believed that the whole cosmos was centred on the Earth (in a geocentric universe). Normally, the planets move steadily from west to east against the background stars. Occasionally, however, they reverse their apparent motion and move from east to west. They retrograde for a period. Then they resume direct motion.

The result of this behaviour was a highly complex pattern, which proved difficult to explain on a geocentric model.

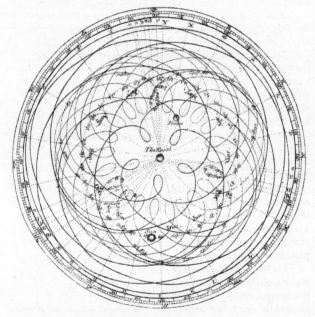

The highly complex pattern of planetary motions that had to be explained on the geocentric model.

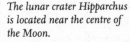

The lunar crater Hipparchus is located near the centre of the Moon.

Because of the notion that movements in the planetary realm could only occur in 'perfect' circles, the idea was introduced that the planets moved on small circles (epicycles) that were themselves carried round the Earth on circular orbits. This concept was first introduced by Apollonius of Perga, whose dates are unknown, but who lived around 240 to 190 BCE. He studied geometry and astronomy, but most of his writings are lost. The crater Apollonius on the Moon carries his name. The various concepts were developed by the Greek astronomer Hipparchus (about 190 to 120 BCE), who also has a lunar crater named after him.

Apart from 'explaining' the retrograde motion of the planets, the epicyclic theory also provided a solution to the apparent changes in the distances of the planets from the Earth.

This way of explaining the motion of the planets, where the circular epicycle was carried around the Earth in a larger circlar orbit prevailed for some years.

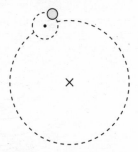

The planet was carried around a small circle, itself carried round the Earth on a larger circle.

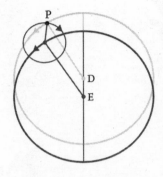

The planet (P) orbited on a small circle, carried round a larger circle, centred at the deferent (D), offset from the Earth (E).

But even this idea proved inadequate to describe the motion of the planets, so the concept of the deferent was added. In this the circular motion of the epicycle was carried round the Earth on a circle that was itself not centred on the Earth, but offset from its centre.

The use of the epicycle and deferent was developed and propagated by the great astronomer Ptolemy (approximate dates 100 to 170 CE), who found that he had to introduce further terms, which he denoted the 'eccentric' and the 'equant'. This further complicated the situation, and these terms are not explained here.

All this complexity became redundant, of course, as soon as the view of Copernicus prevailed in which the Earth itself orbited the Sun. It became apparent that the planets displayed retrograde motion when the Earth 'caught up' and 'passed' the planets in their orbits.

The astronomer Ptolemy, assisted by Urania, the Muse of Astronomy. Ptolemy is shown with a crown because at the time of this image (1508) he was confused with the rulers of Egypt.

February

February – Introduction

There are just two, relatively minor, meteor showers in February 2023, which is a quiet month for astronomers with few notable events. The two showers are both southern ones, and neither is visible to northern observers. The most significant shower is the *Centaurids*, which actually begins on 31 January 2023. This shower has two separate branches, with radiants lying near the brightest stars in Centaurus (*Rigil Kentaurus* and *Hadar*, or α and β Cen, respectively). These two streams are thus known as the *α-Centaurids* and the *β-Centaurids*. Both branches of this shower reach a low maximum, with an hourly rate of about 5–6 meteors per hour, on February 8. That day the Moon is at Day 17 of the lunation, just two days after Full Moon, so observing conditions are not favourable.

The second southern shower, the *γ-Normids*, begins its activity around 25 February 2023 and continues into March, reaching a weak (but very sharp) maximum on March 14–15. Unfortunately, its meteors are difficult to differentiate from sporadics, so are likely to be identified by dedicated meteor observers only. Observing conditions at maximum are not very favourable in 2023, because the Moon is near or at Last Quarter (Days 22 and 23 of the lunation).

The planets
Mercury is essentially invisible in twilight. *Venus* is bright (mag. -3.9) and will become visible in the evening sky towards the end of the month. *Mars* is in *Taurus* and fades from mag. -0.3 to mag. 0.4 over the month. *Jupiter*, in *Pisces*, is mag. -2.1,

Why is February short?
Why does February have such an odd number of days, and why do we tinker with it every four years? The answer is suprisingly complicated, and involves the ancient Roman lunar calendar, Roman emperors, including Julius Caesar, the Roman Senate, the priests, and the way in which politicians messed about with the calendar, and how we have avoided even greater confusion. A fairly comprehensive description of how these changes came about is given on pages 62–63.

but **Saturn** is in **Aquarius**, much closer to the Sun, and will be lost in twilight. **Uranus** remains in **Aries** at mag. 5.8, and **Neptune** is in **Pisces** at mag. 7.9 to 8.0, close to evening twilight.

F

February 2021 was a busy month at Mars. On February 9, a spaceprobe launched by the United Arab Emirates, named **Hope**, arrived to carry out atmospheric studies. The next day, the Chinese **Tianwen-1** mission arrived. This mission consists of an orbiter, a lander and a rover. The lander and rover landed on 14 May 2021 and the **Zhurong** lander was deployed on 22 May 2021. On 18 February 2021, the NASA Perseverance rover, with its helicopter **Ingenuity** was successfully landed in Jezero crater.

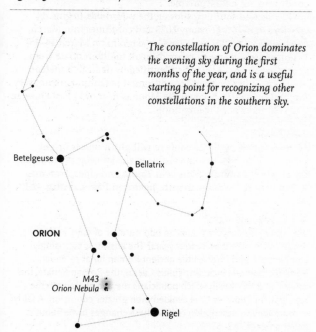

The constellation of Orion dominates the evening sky during the first months of the year, and is a useful starting point for recognizing other constellations in the southern sky.

Betelgeuse

Bellatrix

ORION

M43
Orion Nebula

Rigel

The names and lengths of the months

Several modern names for months derive from the original Roman 'calendar of Romulus', which contained just ten months. The winter was considered to be without Full Moons and thus months. The calendar dates back to around 750 BCE. The legendary king Romulus was supposed to have started naming the months, beginning at the spring equinox (in March). So we have:

- March from **Martius** (Mars, the god of war)
- April from **Aprilis** (the reason for this one is uncertain, and possibly related to the raising of hogs)
- May from **Maius** (a local Italian goddess)
- June from **Junius** (the queen of the Latin gods).

After Junius, he simply counted the months beginning **Quintilis**, **Sextilis** (our July and August). This gave:

- September from **septem** (seven)
- October from **octo** (eight)
- November from **novem** (nine)
- December from **decem** (ten)

About 713 BCE, two more, January (**Januarius** from Janus, the Roman god of beginnings) and February (**Februarius**, from Februa, the Roman ritual of purification at the end of the year) were added.

Originally, the Romans used a lunar calendar, beginning at the spring equinox in March. To keep the lunar calendar in step with the year, what are called intercalary months were sometimes inserted.

The calendar was initially under the control of priests and they fiddled things to their advantage. The calendar was used to determine when people could carry out business, and when ceremonies could occur. The priests added intercalary months whenever they thought fit, for example when more taxes were required. There was utter chaos and the calendar never agreed with the solar year.

Eventually, Julius Caesar instigated the Julian calendar reform. Months alternated between 30 and 31 days. This was slightly too long, giving 366 days, so one day was removed

F

from the last month of the year: February, to give it 29 days. An additional day was returned to February every four years, to keep things in step with the Sun. One month, Quintilis, was named Julius (the month of Caesar's birth).

But then the priests messed things up again. They started counting leap years every three years. The error was corrected by the emperor Augustus and by 8 CE the matter had been solved and the months and the Sun were in agreement. But then the Senate decided to rename one month in honour of Augustus – so the month of Sextilis became our August. Unfortunately, under Caesar's scheme that month had just 30 days, whereas Caesar's (our July) had 31 days. Obviously Augustus had to have the same number of days, so they pinched one from poor February, leaving it with 28 days, except in leap years. (At the same time, to avoid having three months with 31 days in succession they also tinkered with the lengths of the months after August, which is why September and November now have 30 days and October and December 31.

Two later emperors tried to change things again. Domitian (emperor between 81 and 96 CE) wanted to rename September, the month when he became emperor, to *Gemanicus* to commemorate his victory over the German tribes, and then rename October (when he was born) to *Domitianus*. However, the Roman Senators did not approve of Domitian, who had drastically reduced their powers, so they retained the changes initiated by Augustus. Even worse might yet have ensued. A later emperor was Commodus (emperor 176–192 CE), the son of emperor Marcus Aurelius. Towards the end of his reign, he became a megalomaniac, adopting various names. After a disastrous fire devastated Rome in 191 CE, early in 192 Commodus declared himself the new Romulus, and ritually re-founded Rome, renaming the city *Colonia Lucia Annia Commodiana*. He then tried to rename all the months of the year to correspond exactly with his own (now twelve) names: *Lucius*, *Aelius*, *Aurelius*, *Commodus*, *Augustus*, *Herculeus*, *Romanus*, *Exsuperatorius*, *Amazonius*, *Invictus*, *Felix*, and *Pius*.

Sunrise and sunset

City	Date	Sunrise	Sunset
Buenos Aires, Argentina			
	Feb. 01	09:13	23:00
	Feb. 28	09:40	22:32
Cape Town, South Africa			
	Feb. 01	04:07	17:52
	Feb. 28	04:33	17:24
London, UK			
	Feb. 01	07:40	16:50
	Feb. 28	06:49	17:39
Los Angeles, USA			
	Feb. 01	14:50	01:22
	Feb. 28	14:23	01:48
Nairobi, Kenya			
	Feb. 01	03:41	15:51
	Feb. 28	03:41	15:49
Sydney, Australia			
	Feb. 01	19:17	09:01
	Feb. 28	19:42	08:33
Tokyo, Japan			
	Feb. 01	21:41	08:08
	Feb. 28	21:12	08:34
Washington, DC, USA			
	Feb. 01	12:15	22:29
	Feb. 28	11:42	23:00
Wellington, New Zealand			
	Feb. 01	17:27	07:42
	Feb. 28	18:01	07:07

NB: the times given are in Universal Time (UT)

The Moon's phases and ages

Northern hemisphere

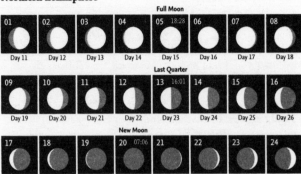

Full Moon

01	02	03	04	05 18:28	06	07	08
Day 11	Day 12	Day 13	Day 14	Day 15	Day 16	Day 17	Day 18

Last Quarter

09	10	11	12	13 16:01	14	15	16
Day 19	Day 20	Day 21	Day 22	Day 23	Day 24	Day 25	Day 26

New Moon

17	18	19	20 07:06	21	22	23	24
Day 27	Day 28	Day 29	Day 30	Day 1	Day 2	Day 3	Day 4

First Quarter

25	26	27 08:06	28
Day 5	Day 6	Day 7	Day 8

Southern hemisphere

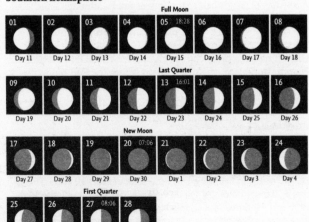

Full Moon

01	02	03	04	05 18:28	06	07	08
Day 11	Day 12	Day 13	Day 14	Day 15	Day 16	Day 17	Day 18

Last Quarter

09	10	11	12	13 16:01	14	15	16
Day 19	Day 20	Day 21	Day 22	Day 23	Day 24	Day 25	Day 26

New Moon

17	18	19	20 07:06	21	22	23	24
Day 27	Day 28	Day 29	Day 30	Day 1	Day 2	Day 3	Day 4

First Quarter

25	26	27 08:06	28
Day 5	Day 6	Day 7	Day 8

F

The Moon

The Moon in February

On February 3, two days before Full Moon, the Moon passes 1.9° south of *Pollux* in *Gemini*. One day after Full Moon, it is 4.5° north of *Regulus* in *Leo*. On February 11, it is 3.5° north of *Spica* in *Virgo*. On February 14, one day after Last Quarter, the Moon is 1.8° north of *Antares* in *Scorpius*. On February 19, one day before New Moon, it is 3.7° south of *Saturn* in *Aquarius*, but this will be lost in twilight. The same problem will apply on February 21, when the Moon is 2.5° south of *Neptune*, which is faint at mag. 7.9. The next day, the Moon is 2.1° south of Venus, which is mag. -3.9, so will be visible in the evening twilight. Later the same day, the Moon is 1.2° south of *Jupiter* (mag. -2.1) in *Pisces*. On February 28, the Moon is 1.1° north of *Mars*, in *Taurus*.

Snow Moon

In the northern hemisphere, February is often the coldest month, and most countries on both sides of the Atlantic see significant falls of snow. The Full Moon of February is thus often called the 'Snow Moon', although just occasionally that name has been applied to the Full Moon in January. Some North American tribes named it the 'Hunger Moon' because of the scarcity of food sources during the depths of winter, while other names are 'Storm Moon' and 'Chaste Moon', although the last name is more commonly applied to the Full Moon in March. To the Arapaho of the Great Plains, the Full Moon was called the Moon 'when snow blows like grain in the wind'.

Black Moon

Because it never has more than 29 days, and the synodic month (between any pair of phases, such as from New Moon to New Moon) is slightly more than half a day longer, sometimes there is no Full Moon in February. This occurs about every 19 years. This is one of the definitions of the term '*Black Moon*', although the same term is sometimes applied to a New Moon that does not occur in a particular season, as reckoned from the equinoxes or solstices.

F

The Chelyabinsk meteorite

At approximately 03:30 UT on 15 February 2013 a bolide entered the Earth's atmosphere above the city of Chelyabinsk in the southern Urals of Russia. The airburst, at a height of about 29.7 km, was extremely bright – brighter than the Sun, and visible over 100 km away. The shockwave arrived at the ground some seconds later and caused extensive damage, with some 1500 people injured, mainly from glass broken by the shockwave.

The initial object (a small asteroid) was estimated to have a diameter of about 20 metres and approached from the direction of the Sun, which is why it was not detected before arrival. Following the airburst, some fragments survived to reach the ground, west of Chelyabinsk. The snow-covered ground made the recovery of these relatively easy. Most, however, found their way into private hands. The largest known fragment was eventually recovered from the frozen Lake Cherbakul after a long recovery process. This proved to have an initial mass of 654 kg. All the recovered meteorites were found to be ordinary stony chondrites (see pages 74–75).

A fragment of the meteorite recovered shortly after the Chelyabinsk fall. This fragment weighs 112.2 g. It displays a fusion crust and a highly shocked, chondritic interior. The measurement cube is 1 cm on a side.

Calendar for February

03	20:25	Pollux 1.9°N of the Moon
04	08:55	Moon at apogee = 406,476 km
05	18:28	Full Moon
06	18:22	Regulus 4.5°S of the Moon
08		α-Centaurid meteor shower maximum
11	05:02	Spica 3.5°S of the Moon
13	16:01	Last Quarter
14	18:43	Antares 1.8°S of the Moon
15	12:00 *	Neptune (mag. 8.0) 0.01°N of Venus (mag. -3.9)
16	16:48	Saturn at superior conjunction
18	20:52	Mercury 3.6°N of the Moon
19	09:06	Moon at perigee = 358,267 km
19	23:58	Saturn 3.7°N of the Moon
20	07:06	New Moon
21	18:16	Neptune 2.5°N of the Moon
22	07:55	Venus 2.1°N of the Moon
22	22:00	Jupiter 1.2°N of the Moon
25–Mar.28		γ-Normid meteor shower
25	13:05	Uranus 1.3°S of the Moon
27	08:06	First Quarter
27	13:39	Aldebaran 8.5°S of the Moon
28	04:32	Mars 1.1°S of the Moon

These objects are close together for an extended period around this time.

February 3 • *The Moon forms a nice, almost isosceles, triangle with Pollux and Castor (as seen from Sydney).*

February 14–15 • *The Moon passes Antares. The 'Cat's Eyes' are close to the horizon (as seen from central USA).*

February 22 • *The crescent Moon with Venus and Jupiter. Diphda (β Cet) is closer to the horizon (as seen from London).*

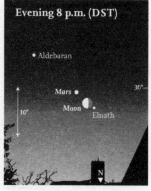

February 28 • *In the north, the Moon is lining up with Elnath (β Tau), Mars, and Aldebaran. (as seen from Sydney).*

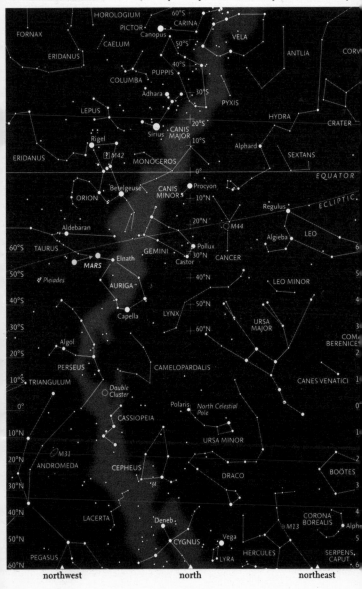

February – Looking North

The months of January and February are probably the best time for seeing the section of the Milky Way that runs in the northern and western sky from **Orion** and **Gemini** right through **Auriga**, **Perseus** and **Cassiopeia**, towards **Cygnus**, low in the north. Although not as readily visible as the denser star clouds of the summer Milky Way, on a clear night so many stars may be seen that even a distinctive constellation such as Cassiopeia, which lies across the Milky Way, is not immediately obvious.

The 'base' of the constellation of **Cepheus** lies on the edge of the stars of the Milky Way, but the red supergiant star **Mu** (μ) **Cephei**, called the 'Garnet Star' by William Herschel with its striking red colour remains readily visible. The groups of stars, known as the **Double Cluster** in Perseus (NGC 869 & NGC 884, often known as h and χ Persei), lying between Perseus and Cassiopeia, are well-placed for observation.

Beyond the Milky Way, Perseus and Cassiopeia, the constellation of **Andromeda** is beginning to be lost in the north-western sky.

Castor and **Pollux** in **Gemini** are near the zenith for observers at 30°N, but for most northern observers the constellation is best seen when facing south. The head of **Draco** is now higher in the sky and easier to recognize, with its long, straggling 'body' curling round **Ursa Minor** and the Pole. **Ursa Major** has begun to swing round towards the north, and Cassiopeia is now lower in the western sky.

For observers in the far north, most of **Cygnus**, with its brightest star, **Deneb** (α Cygni), is visible in the north, and even **Lyra**, with **Vega** (α Lyrae) may be seen at times. Most of the constellation of **Hercules** is visible, together with the distinctive circlet of **Corona Borealis** to its east. Observers farther south may see Deneb and even Vega peeping over the northern horizon at times during the night, although they will often be lost (like all the fainter stars) in the inevitable extinction along the horizon.

The Japanese spacecraft **Hayabusa 2** obtained the first sample of asteroid **Ryugu** on 21 February 2019.

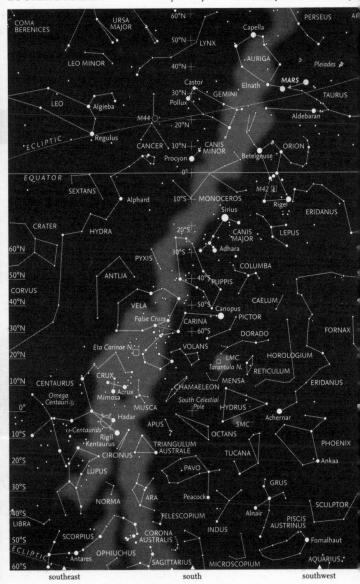

southeast   south   southwest

72

February – Looking South

Orion is now beginning to sink into the southwest, and the two brightest stars in the sky, **Sirius** and **Canopus** (α Carinae), are readily visible to observers at low northern latitudes and, of course, to those who are south of the equator. (Canopus is close to the zenith for those in the far south.)

South of **Carina** and the neighbouring constellation of **Vela** (both part of the original, and now obsolete, constellation of Argo Navis), lies the sprawling constellation of **Centaurus**, surrounding the distinctive constellation of **Crux**, the Southern Cross, which is on the horizon at 10°N. North of this, the 'False Cross', sometimes mistaken for the true constellation of Crux, consists of two stars from each of Vela (δ and κ Velorum) and Carina (ε and ι Carinae). The two brightest stars of Centaurus, **Rigil Kentaurus** (α Centauri) and **Hadar** (β Centauri) are slightly farther south, beyond Crux.

Sirius, α Canis Majoris (α Cma), in the southern celestial hemisphere, is the brightest star in the sky at magnitude -1.44.

The **Large Magellanic Cloud** (LMC) is almost on the meridian, early in the night, to the west of Carina and the small constellation of **Volans**. It is surrounded by several small constellations: **Mensa**, **Reticulum** and **Dorado**. (Technically, it is largely within the area of Dorado.) Any optical aid (such as binoculars) will begin to show some of the remarkable structures within the LMC, including the great **Tarantula Nebula**, or **30 Doradus**, a site of active star formation.

Farther south and west lies **Achernar** (α Eridani), the bright star at the end of **Eridanus**, the long straggling constellation that represents a river and that may now be traced all the way from where it begins near **Rigel** (β Orionis) in Orion.

Beyond Crux, and on the other side of the Milky Way, lies the rest of Centaurus. Northeast of Crux is the finest and brightest globular cluster in the sky, **Omega** (ω) **Centauri** also known as NGC 5139. It is the largest globular cluster in our Galaxy and is estimated to contain about 10 million stars. Although appearing like a star, its non-stellar nature was discovered by Edmond Halley in 1677.

Meteorites

At 21:54 on the evening of 28 February 2021, a brilliant fireball (a bolide) was observed over Gloucestershire. Observations allowed the orbit of the parent body to be determined, showing that the body's original location was the outer region of the asteroid belt between Mars and Jupiter.

Meteorites are conventionally divided into 'falls', where the body is seen to fall and the location is therefore known (at least approximately) and 'finds', where the object is merely found by chance. There are, of course, far more finds than falls, and the numbers in collections are about 65,000 and 1500, respectively.

Collected meteorites may be grouped into three broad classes: irons, stones and stony-irons. The irons tend to be the most distinctive in appearance, but the stones and stony-irons resemble many Earth rocks. Because of their nature, iron meteorites are most likely to survive passage through the Earth's atmosphere relatively intact and create impact craters on the surface. The most famous such impact crater is possibly the Barringer crater in Arizona, which is associated with the 'Canyon Diablo Irons'.

The most numerous class is that of the stones, and these largely consist of silicaceous material. One striking feature of some bodies is the presence of numerous chondrules. Meteorites with these are known, unsurprisingly, as chondrites. Chondrules are small, spherical bodies of silicate minerals that appear to have been melted and formed when floating in space. Their age is estimated at 4.55 thousand million years and they are believed to be material that never condensed into larger bodies. Chondrites (like the comets) are thus thought to be some of the very oldest objects in the Solar System.

Of the chondrites, the most important scientifically are the carbonaceous chondrites, like the Winchcombe meteorite. Apart from the chondrules, these may contain organic material (such as amino acids), water and pre-solar grains. These compounds are essential for life, and there is an opinion that life on Earth has arisen because these materials have been delivered by carbonaceous meteorites or carbonaceous minor planets that have impacted on the Earth.

The main Winchcombe meteorite, recovered from a driveway in the market town.

F

Some carbonaceous chondrites contain very high percentages of water – 'high' implying between 3 and 22 per cent. Many show evidence of being considerably altered by the presence of liquid water.

The most famous carbonaceous chondrite meteorite is probably the Murchison meteorite, observed to fall near Murchison, Victoria, Australia on 28 September 1969. Both Murchison and the recent Winchcombe meteorite belong to the group known as the CM meteorites. Their material appears to resemble that collected by the *Hayabusa 2* spaceprobe from the minor planet *(162173) Ryugu*, and returned to Earth in December 2020. The Murchison meteorite has been particularly important and the subject of numerous, significant studies. It has been found to contain a phenomenal number of molecular compounds (at least 14,000), including some 70 amino acids. Some estimates put the number of potential compounds in the meteorite at hundreds of thousands, or even as high as one million. In January 2020, an international team of cosmochemists announced that some silicon carbide (SiC, carborundum) particles from the Murchison meteorite were the very oldest particles ever detected. They had anomalous isotopic ratios of silicon and carbon, implying that they were formed outside the Solar System. These grains have a suggested age of 7000 million years, some 2500 million years older than the Solar System itself.

March

March – Introduction

The Sun crosses the celestial equator from south to north at the vernal equinox, on Monday, March 20 in 2023. Day and night are then of almost equal length, although refraction in the atmosphere 'raises' the Sun above the horizon at its rising and setting, so daylight at the equinox itself is slightly longer than darkness. (The hours of daylight and darkness change most rapidly around the equinoxes in March and September.) The northern season of spring is considered to begin at the equinox and ancient calendars started the year at that date. (Until Pope Gregory revised the calendar in 1582.) It is also in March that Daylight Saving Time (DST), known in Britain as British Summer Time (BST), begins (on Saturday–Sunday, March 25–26). (In Europe, Daylight Saving Time is introduced on the same date.)

The point at which the Sun crosses the celestial equator is known as the *First Point of Aries*, and originally lay in that constellation. Because of the phenomenon of precession, which changes the orientation of the Earth's axis in space, this point now actually lies in the constellation of *Pisces*, close to the border with Aquarius, lying near the star λ Piscium. It is slowly moving towards Aquarius at a rate of about one degree every 70 years.

The autumnal equinox (or spring equinox for those in the southern hemisphere), when the Sun crosses the celestial equator in the opposite direction (north to south) occurs

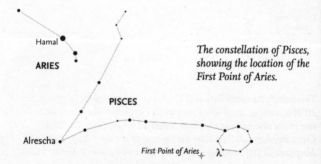

The constellation of Pisces, showing the location of the First Point of Aries.

in September (September 23 in 2023). It lies in the large constellation of **Virgo** (see page 176).

The First Point of Aries is sometimes known as the Cusp of Aries, and the equinox in September as the Cusp of Virgo.

The planets

Mercury is too close to the Sun to be visible this month. It reaches superior conjunction, on the far side of the Sun, on March 17. **Venus**, in the evening sky, is very bright (mag. -3.9 to -4.0), but too close to the Sun to be readily seen. *Mars* is initially at magnitude 0.4 in **Taurus**, but moves into **Gemini** and fades to mag. 1.0. *Jupiter* is in **Pisces**, but is too close to the Sun to be readily visible this month. *Saturn* is in **Aquarius** and lies too far into the morning twilight to be seen. *Uranus* is in **Aries** at mag. 5.8 and **Neptune** (mag. 8.0) is in **Pisces**. That planet comes to superior conjunction on March 15.

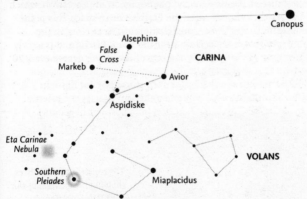

This month the constellation of Carina is well placed for observers at 20°N or more to the south. Canopus (α Car) is the second brightest star in the sky. Avior (ε Car) and Aspidiske (ι Car) form the False Cross, together with two stars that belong to the constellation of Vela: Alsephina (δ Vel) and Markeb (κ Vel). The small constellation of Volans is almost embedded in Carina.

The Equinoxes

At the equinox in March and again, in September, the Sun rises due east and sets due west. In theory, but not in practice, day and night are of equal length (see page 78). At the equinoxes, the Earth's axis is exactly at right angles to the Earth–Sun line, and the Earth is neither tilted towards, nor away from the Sun.

Most early civilizations were in the northern hemisphere and they started their year, and the cycle of seasons, at the spring equinox. When Julius Caesar established what is known as the Julian calendar (page 62) he set the equinox at 25 March, and the year started on that date. Because a year in the Julian calendar is slightly longer than the year as determined by the Sun, the date of the equinox slowly became earlier and by 1580 had drifted back to 21 March. Pope Gregory XIII instigated the reforms that led to what is now known as the Gregorian calendar, which he introduced in October 1582, and which we still use. Even so, there is a slight variation in the exact date of the equinox, because this, of course, is astronomically defined. The calendar date varies from March 19 to March 21 over the course of the 400-year long leap-year cycle. The earliest equinox in the twenty-first century will be 19 March 2096, and the latest was 21 March 2003. There are, of course, corresponding differences in the date of the northern autumnal (southern spring) equinox in September; the earliest being 21 September 2096 and the latest 23 September 2003.

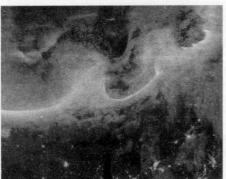

Part of the auroral oval, photographed over Canada from the International Space Station. Parts of the Great Lakes may be seen at the bottom of the picture.

Effects of the equinoxes

There are three different effects that common ideas associate
with the equinoxes: One meteorological, and two astronomical.
There are strong beliefs in 'equinoctial gales', 'equinoctial tides'
and 'equinoctial aurorae'. It is often thought that gales are more
frequent and strongest at the equinoxes, particularly at the
(northern) autumnal equinox. In fact, there are most gale-force
winds around the time of the winter solstice in late December
and early January. The concept has probably arisen because
after the quiet period of summer, depressions, with their
accompanying winds, tend to move south and bring high winds
to the British Isles.

There are indeed equinoctial tides, and these tend to be
greater in extent than at other times of the year. The effect is
well-known and was explained by Isaac Newton in the late-
seventeenth century. He explained the effect by stating that the
gravitational effect of the Sun is greatest when its declination
(page 8) is at a minimum. It is then acting along a line to the
centre of the Earth, and is not offset to north or south. The Sun's
declination is precisely zero when it crosses the celestial equator,
and this occurs, twice a year, at the equinoxes. At the equinoxes,
the Sun is directly over the Earth's equator, and thus exerts the
greatest influence over the tides. However, the effect is greatest
at the equator itself, and is often masked by other effects, such as
those caused by atmospheric pressure and winds.

When it comes to equinoctial aurorae, there is definitely
an increase in the frequency of aurorae and geomagnetic
storms. This is known as the Russell–McPherron effect. The
Sun's magnetic field extends out into interplanetary space,
particularly at the poles. Closer to the Sun's equator, the
magnetic-field lines form closed loops, rejoining the surface.
At certain times of the year – which happen to coincide with
the equinoxes – the north or south poles are tilted towards
the Earth. (The south pole in March and the north pole in
September.) The Sun's magnetic field is able to connect directly
to that of the Earth, so charged particles from the Sun are
able to cascade down into the atmosphere and create more
geomagnetic storms and aurorae.

Sunrise and sunset

City	Date	Sunrise	Sunset
Buenos Aires, Argentina			
	Mar. 01	09:41	22:30
	Mar. 31	10:05	21:50
Cape Town, South Africa			
	Mar. 01	04:34	17:23
	Mar. 31	04:57	16:43
London, UK			
	Mar. 01	06:47	17:41
	Mar. 31	05:39	18:32
Los Angeles, USA			
	Mar. 01	14:22	01:49
	Mar. 31	13:42	02:12
Nairobi, Kenya			
	Mar. 01	03:41	15:49
	Mar. 31	03:34	15:40
Sydney, Australia			
	Mar. 01	19:43	08:32
	Mar. 31	20:07	07:52
Tokyo, Japan			
	Mar. 01	21:11	08:35
	Mar. 31	20:29	09:01
Washington, DC, USA			
	Mar. 01	11:41	23:01
	Mar. 31	10:55	23:31
Wellington, New Zealand			
	Mar. 01	18:02	07:05
	Mar. 31	18:36	06:15

NB: the times given are in Universal Time (UT)

The Moon's phases and ages

Northern hemisphere

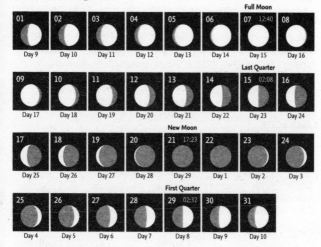

Southern hemisphere

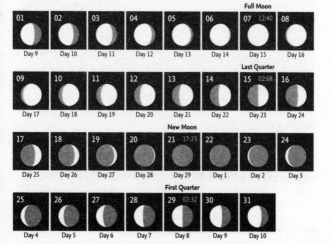

The Moon

The Moon in March

On March 3, the waxing gibbous Moon is 1.7° south of **Pollux** (mag. 1.1) the brightest star in **Gemini**. On March 6, one day before Full Moon, it is 4.5° north of **Regulus** (mag. 1.4) in **Leo**. By March 10 it is 3.4° north of **Spica** and by March 14, one day before Last Quarter, it is 1.6° north of **Antares** in **Scorpius**. On March 19, the waning crescent is 3.6° south of **Saturn** (mag. 0.8). At New Moon, on March 21, it is 2.4° south of faint **Neptune** (mag. 0.8). The next day in twilight, it passes south of **Mercury**, and then 0.5° south of **Jupiter** (mag. -2.1). On March 24, the Moon occults **Venus**, partly visible from southeast Asia. On March 26, the waxing crescent is 8.7° north of **Aldebaran** in **Taurus**. It passes 2.3° north of **Mars** and on March 30 is again 1.6° south of **Pollux**.

Worm Moon

One name for the last Full Moon of the winter season, which falls in March, is the 'Worm Moon'. The name derives from the fact that earthworms become active in the soil at the end of winter and are sometimes seen at the surface. Other names include 'Crow Moon', because the birds become particularly active and are avid to feed on the worms, after the lack of food during the winter months. Another name is the 'Sap Moon', which is particularly relevant in Canada, because this is the time when maple trees may be tapped for the sap (to produce maple syrup, beloved by Canadians). In Europe, the term 'Lenten Moon' was sometimes used, and this is the Old English/Anglo-Saxon name for this particular Full Moon, and is derived, of course, from the Christian period of Lent.

Ceres

Ceres was originally discovered by Giuseppe Piazzi on 1 January 1801 at Palermo Observatory in Sicily. Although originally classed as a planet, orbiting between the orbits of Mars and Jupiter, it was reclassified as a dwarf planet in 2006. It is now known to have a diameter of 939 km.

When the second such body, (2) Pallas, was discovered in 1802, William Herschel proposed the name 'asteroid' for these objects, because of their star-like appearance. The term 'minor planet'

An image of Ceres, obtained by the Dawn spacecraft on 4 May 2015 from a distance of 13,641 km. Ceres comes to opposition on 21 March 2023.

has also been extensively used for these bodies. Both terms are found in modern literature.

On very rare occasions, under very dark skies, it is just possible to make out the dwarf planet with the naked eye, when it reaches magnitude 6.7. Generally, however, it requires telescopic or binocular aid to become visible. At opposition in 2023 it will reach magnitude 7.4. It did not come to opposition in 2022.

Ceres was visited by the NASA spaceprobe **Dawn**, which went into orbit around the body in 2015. Dawn had previously orbited the minor planet (asteroid) (4) Vesta, which it had circled for 13 months. Dawn carried out numerous scientific experiments at Ceres and mapped the surface in great detail. The surface was found to consist of water ice and carbonate and clay minerals. Gravity data suggest that the centre consists of an icy rock-mantle and core, with perhaps a frozen icy crust. Certain unfrozen heavy brines may occasionally reach the surface and produce 'cryptovolcanism' (features created by frozen ices).

The Dawn mission came to an end on 1 November 2018, when the spaceprobe exhausted all its fuel.

The path of minor planet (1) Ceres, around its opposition on March 21. Stars are shown down to magnitude 8.0.

Calendar for March

02	10:00 *	Saturn (mag. 0.9) 0.9°N of Mercury (mag. -0.6)
02	11:00 *	Jupiter (mag. -2.1) 0.5°S of Venus (mag. -3.9)
03	02:48	Pollux 1.7°N of the Moon
03	15:00	Moon at apogee = 405,889 km
06	00:46	Regulus 4.5°S of the Moon
07	12:40	Full Moon
10	10:45	Spica 3.4°S of the Moon
14	00:56	Antares 1.6°S of the Moon
15		γ-Normid meteor shower maximum
15	02:08	Last Quarter
15	23:39	Neptune at superior conjunction
16	15:00 *	Neptune (mag. 8.0) 0.4°N of Mercury (mag. -1.8)
17	10:45	Mercury at superior conjunction
19	15:12	Moon at perigee = 362,697 km
19	15:22	Saturn 3.6°N of the Moon
20	21:24	March Equinox
21	06:47	Neptune 2.4°N of the Moon
21	07:41	Dwarf planet Ceres at opposition (mag. 6.9)
21	17:23	New Moon
22	00:10	Mercury 1.8°N of the Moon
22	19:56	Jupiter 0.5°N of the Moon
24	10:26	Venus occulted by the Moon
25	00:39	Uranus 1.5°S of the Moon
26	01:00	European Daylight Saving Time) begins
26	22:05	Aldebaran 8.7°S of the Moon
28	13:16	Mars 2.3°S of the Moon
28	15:00 *	Mercury (mag. -1.4) 1.5°N of Jupiter (mag. -2.1)
29	02:32	First Quarter
30	10:02	Pollux 1.6°N of the Moon
31	06:00 *	Uranus (mag. 5.8) 1.3°S of Venus (mag. -4.0)
31	11:17	Moon at apogee = 404,919 km

These objects are close together for an extended period around this time.

March 2 • *After sunset, Venus (mag. -3.9) and Jupiter (mag. -2.1) are close together in the western sky (as seen from London).*

March 22–24 • *In the evening twilight, the narrow crescent Moon passes Jupiter and Venus (as seen from central USA).*

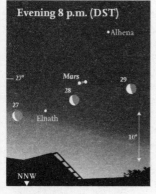

March 27–29 • *The waxing crescent Moon passes Elnath (β Tau) and Mars (as seen from Sydney).*

March 30 • *In the early morning, the Moon forms a nice triangle with Castor and Pollux (as seen from London).*

northwest   north   northeast

March – Looking North

The highly distinctive (and widely recognized) constellation of **Ursa Major** with the distinctive asterism of the Plough (or Big Dipper) is now 'upside down' and near the zenith for observers in the far north, for whom it is particularly difficult to observe. At this time of year, it is high in the sky for anyone north of the equator. Only observers farther towards the south will find it lower down towards their northern horizon and reasonably easy to see. However, at 30°S, even the seven stars making up the main, easily recognized portion of the constellation are too low to be visible.

Auriga, with brilliant **Capella** (α Aurigae) is also very high on the opposite side of the meridian. The constellation of **Perseus** lies between it and **Andromeda** on the western side of the sky.

Ursa Minor, also with seven main stars, one of which is **Polaris**, the Pole Star, and the long constellation of **Draco** that winds around the Pole, are readily visible for anyone in the northern hemisphere, although, of course, Polaris is right on the horizon for anyone at the equator, and thus always lost to sight. **Cepheus** is near the meridian to the north, with **Cassiopeia**, to its west beginning to turn and resume its 'W' shape. The constellation of Andromeda is now diving down into the northwestern sky. In the east, beyond **Alkaid** (η Ursae Majoris), the final star in the 'tail' of Ursa Major, lies the top of **Boötes**. Farther to the south, most of **Hercules** and the 'Keystone' shape that forms the major portion of the body is visible.

On 13 March 1989 a major geomagnetic storm created a nine-hour disruption of Hydro-Quebec's electricity transmission system. The accompanying aurorae could be seen as far south as Texas and Florida. The geomagnetic storm was one of a number of incidents during a phase of major solar activity.

Observers at 50°N may occasionally be able to detect bright **Deneb** (α Cygni) and **Vega** (α Lyrae) skimming the horizon, together with portions of those particular constellations, although most of the time they will be lost in the extinction that occurs at such low altitudes.

southeast   south   southwest

March – Looking South

The distinctive constellation of *Leo* is close to the meridian early in the night, clearly visible for anyone north of the equator. It is easily recognized, with bright *Regulus* (α Leonis) at the base of the 'backward question mark' of the asterism known as 'the Sickle', which lies north of Regulus. To the west of Leo is the inconspicuous constellation of *Cancer*, and still farther away from the meridian, the far more striking constellation of *Gemini*, with the bright stars *Castor* (α Geminorum), the closer to the Pole, and *Pollux* (β Geminorum). The constellation straddles the ecliptic, and Pollux may sometimes be occulted by the Moon (as may Regulus), although no such occultation occurs in 2023. Castor is remarkable in that even a fairly small telescope will show it as consisting of three stars (two fairly bright, and one fainter). However, more detailed investigation reveals that each of those stars is actually a double, so the whole system consists of no fewer than six stars.

Below Cancer is the very distinctive asterism of the 'Head of Hydra', consisting of five (or six) stars, that is the western end of the long constellation of *Hydra*, the largest constellation in the sky, that runs far towards the east, roughly parallel to the ecliptic. *Alphard* (α Hydrae) is south, and slightly to the west of Regulus in Leo and is relatively easy to recognize as it is the only fairly bright star in that region of the sky. North of Hydra and between it and the ecliptic and the constellation of *Virgo* are the two constellations of *Crater* and *Corvus*. Farther west, the small constellation of *Sextans* lies between Hydra and Leo.

Farther south, the Milky Way runs diagonally across the sky, and the constellation of *Vela* straddles the meridian. Slightly farther south is the constellation of *Carina*, with, to the west, brilliant *Canopus* (α Carinae), which lies below the constellation of *Puppis*, which is itself between Vela and *Canis Major* in the west.

Crux (the Southern Cross) is southeast of Carina and the two principal stars of *Centaurus*, *Rigil Kentaurus* (α Centauri) and *Hadar* (β Centauri). The *Large Magellanic Cloud* (LMC) lies west of these stars, on the other side of the meridian.

Easter Sunday

Easter Sunday is calculated to occur after the Full Moon following March 21. Now that astronomers know the phases of the Moon decades or hundreds of years in advance, you might think that knowing the date of Easter would be easy. But it is not as easy as that. The Christian religious festival is calculated using a lunar calendar: the ecclesiastical calendar. In this, the 'months' alternate between 30 and 29 days. To astronomers, the time between New Moon and New Moon (known as the synodic month) averages 29.530 587 981 days (to take it to nine decimal places). That is just slightly more than twenty-nine and a half days, so 'months' that alternate between 29 and 30 days do give a reasonable agreement. (But not always, see the description of the term 'Black Moon' on page 66.) In the ecclesiastical calendar, Full Moon is always taken to be on the 14th day of the lunar month, reckoned in local time. So the ecclesiastical calendar may get out of step with the astronomical calendar, in which Full Moon is defined as occurring at a specific date and time (in Co-ordinated Universal Time, UTC). What is more, to astronomers, the exact date and time of Full Moon apply worldwide. In 2023, the religious Easter Sunday is on April 9 (three days after Full Moon), whereas if calculated astronomically it would be on April 6, at Full Moon.

On 3 March 2022, a spent rocket part crashed into the lunar far side. Although denied by China, this is believed to have been part of the rocket that carried the Chinese lunar lander *Chang-e 4* to the Moon.

April

April – Introduction

The principal astronomical event in April is the hybrid eclipse of April 20, with maximum eclipse over Indonesia. Such hybrid eclipses are unusual, because totality is extremely short. In this case the duration of totality is just 1 minute 16 seconds. Only for this short time is the Moon close enough to the Earth to completely cover the disc of the Sun. Because of the curvature of the Earth, the distance is sufficiently great for the eclipse to appear as annular both before and after totality.

Apart from the eclipse, Mercury comes to greatest eastern elongation on April 11, and may be glimpsed in the evening sky. Three meteor showers are active during the month. Two are best seen from the northern hemisphere, with one significant southern shower. The one moderate, northern shower is the **Lyrids** (often called the **April Lyrids** to distinguish them from several minor showers that originate in the constellation at various times during the year). In 2023 the shower begins on April 14, one day after Last Quarter, and comes to a weak maximum of 18

The charts show the location of the April Lyrids radiant (top), the π-Puppids radiant (middle) and the η-Aquariids radiant (bottom).

meteors per hour on April 22–23, 2–3 days after New Moon, so conditions are generally favourable.

There is another, stronger shower, with a possible hourly rate of 50 meteors per hour. This is the *η-Aquariid* shower. This begins on April 19, the day before New Moon, and peaks on May 6, the day after Full Moon, continuing well into May (May 28). So conditions for observing this shower are good at the beginning, deteriorate around maximum, but then improve again after Full Moon on May 6. The *η-Aquariid* shower, like the Orionids in October, is the result of particles left in orbit behind the famous Comet 1P/Halley. This is a moderately predictable shower, and the maximum rate is between 40 and 50 meteors per hour. The radiant for this shower is close to the 'Y'-shaped asterism known as the 'Water Jar' in *Aquarius*.

The southern meteor shower, the *π-Puppids*, begins in April. This shower starts to be active on April 15, two days after Last Quarter and lasts until April 28, one day after First Quarter, with maximum on the night of April 23–24 when the Moon is a waxing crescent. The shower was unknown until 1972. The rate seems to be variable, reaching a maximum of about 40 meteors per hour in 1977 and 1983. It is difficult to predict how many meteors will be seen. As with the Lyrids, maximum occurs just after New Moon. The parent comet is believed to be Comet 26P/Grigg–Skjellerup.

On 5 April 2019, the Japanese spacecraft *Hayabusa 2* fired an impactor at the asteroid *Ryugu*, creating a crater from which samples were later obtained.

The planets
Mercury is close to the Sun. It passed superior conjunction on March 17, and comes to greatest eastern elongation (19.5° from the Sun, at mag. -0.0) on April 11. *Venus* is very bright (mag. -4.0 to -4.2) in the evening sky. *Mars* (mag. 1.1 to 1.3) moves across *Gemini* over the month. *Jupiter* is lost in the twilight in *Pisces*. *Saturn* (mag. 1.0.) is in *Aquarius*, very low in the morning twilight. *Uranus* (mag. 5.8 to 5.9) is slowly moving eastwards in *Aries*. *Neptune* is still in *Pisces*, at mag. 8.0 to 7.9.

Sunrise and sunset

City	Date	Sunrise	Sunset
Buenos Aires, Argentina			
	Apr. 01	10:06	21:48
	Apr. 30	10:28	21:13
Cape Town, South Africa			
	Apr. 01	04:58	16:42
	Apr. 30	05:20	16:06
London, UK			
	Apr. 01	05:37	18:34
	Apr. 30	04:35	19:22
Los Angeles, USA			
	Apr. 01	13:41	02:13
	Apr. 30	13:05	02:36
Nairobi, Kenya			
	Apr. 01	03:34	15:40
	Apr. 30	03:28	15:32
Sydney, Australia			
	Apr. 01	20:08	07:51
	Apr. 30	20:30	07:15
Tokyo, Japan			
	Apr. 01	20:27	09:02
	Apr. 30	19:50	09:26
Washington, DC, USA			
	Apr. 01	10:53	23:32
	Apr. 30	10:12	24:00
Wellington, New Zealand			
	Apr. 01	18:37	06:14
	Apr. 30	19:07	05:30

NB: *the times given are in Universal Time (UT)*

The Moon's phases and ages

Northern hemisphere

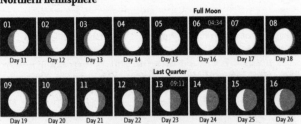

Full Moon

| 01 | 02 | 03 | 04 | 05 | 06 04:34 | 07 | 08 |
| Day 11 | Day 12 | Day 13 | Day 14 | Day 15 | Day 16 | Day 17 | Day 18 |

Last Quarter

| 09 | 10 | 11 | 12 | 13 09:11 | 14 | 15 | 16 |
| Day 19 | Day 20 | Day 21 | Day 22 | Day 23 | Day 24 | Day 25 | Day 26 |

New Moon

| 17 | 18 | 19 | 20 04:12 | 21 | 22 | 23 | 24 |
| Day 27 | Day 28 | Day 29 | Day 30 | Day 1 | Day 2 | Day 3 | Day 4 |

First Quarter

| 25 | 26 | 27 21:20 | 28 | 29 | 30 |
| Day 5 | Day 6 | Day 7 | Day 8 | Day 9 | Day 10 |

Southern hemisphere

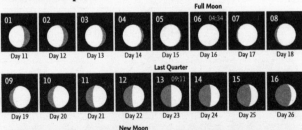

Full Moon

| 01 | 02 | 03 | 04 | 05 | 06 04:34 | 07 | 08 |
| Day 11 | Day 12 | Day 13 | Day 14 | Day 15 | Day 16 | Day 17 | Day 18 |

Last Quarter

| 09 | 10 | 11 | 12 | 13 09:11 | 14 | 15 | 16 |
| Day 19 | Day 20 | Day 21 | Day 22 | Day 23 | Day 24 | Day 25 | Day 26 |

New Moon

| 17 | 18 | 19 | 20 04:12 | 21 | 22 | 23 | 24 |
| Day 27 | Day 28 | Day 29 | Day 30 | Day 1 | Day 2 | Day 3 | Day 4 |

First Quarter

| 25 | 26 | 27 21:20 | 28 | 29 | 30 |
| Day 5 | Day 6 | Day 7 | Day 8 | Day 9 | Day 10 |

A

The Moon

On April 2, the waxing gibbous Moon is 4.6° north of **Regulus** (mag. 1.4). On April 6, just after Full Moon, it is 3.3° north of **Spica** in **Virgo**. By April 10, it is 1.5° north of **Antares** in the morning sky. On April 16, it passes 3.5° south of **Saturn** low in the evening sky. On April 19, one day before New Moon, it is 0.1° north of **Jupiter** in **Pisces**. On April 20, there is a hybrid **eclipse**, visible from Indonesia (described on page 94). The next day, the Moon is 1.9° south of **Mercury** and, later that day, 1.7° north of **Uranus**, both too faint to be readily visible. By April 23 the waxing crescent Moon is 8.8° north of **Aldebaran** in **Taurus**. Later that day it is 1.3° north of brilliant **Venus** (mag. -4.1). On April 26, the Moon is 3.2° north of **Mars** in **Gemini** and then 1.5° south of **Pollux**. By April 29, it is again 4.6° north of **Regulus**, as it was at the beginning of the month.

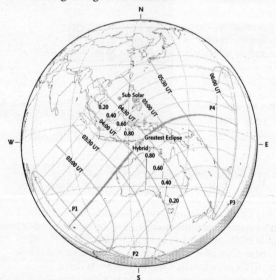

The path of the hybrid eclipse of April 20. For most of the track, the eclipse is annular and only total for a very short period. (See page 94 for more details.)

Pink Moon

The Full Moon in April is known in North America as the Pink Moon, from the pink flowers – phlox – that bloom in the early spring. Other names for this Full Moon used by Native American tribes include Sprouting Grass Moon, Fish Moon, and Hare Moon. On the other side of the Atlantic there is the Old English/Anglo-Saxon name of Egg Moon. In Europe generally it is sometimes known as the Paschal Moon because it is used to calculate the date for Easter (see page 92).

Apollo 13

The ill-fated *Apollo 13* mission was launched from the Kennedy Space Center in Florida on 11 April 1970. The intention was for the spacecraft to be placed into orbit around the Moon and for a landing to take place in the Fra Mauro region, believed to contain material displaced by the impact that created the Mare Imbrium. (The next flight, *Apollo 14*, did successfully land at the Fra Mauro region.) The three-man crew were James Lovell (Commander), John Swigert (Command Module Pilot), and Fred Haise (Lunar Module Pilot). Swigert was to remain in lunar orbit, while Lovell and Haise descended to the surface in the Lunar Module, to carry out their exploration and experiments there.

The interior of Fra Mauro crater, showing the linear elements that are the result of ejecta from the Mare Imbrium impact.

A

In the event, the spacecraft was approximately two-thirds of the way towards the Moon, when on April 14, a routine operation caused one of the two oxygen tanks on the Service Module to rupture, resulting in the loss of oxygen from both tanks. This

Top: *Fra Mauro crater, as photographed from the Apollo 16 mission.*

Right: *The damaged Service Module, photographed shortly after it had been jettisoned, prior to atmospheric re-entry.*

led to the famous statement, initially from Swigert and then Lovell: 'Houston, we've had a problem here ...'. The rupture led, not only to a loss of vital oxygen, but also to a loss of power from the Service Module's fuel cells, which drew oxygen from the two tanks. The Command Module, where the astronauts would normally spend their time, lost power (normally supplied by the Service Module), and oxygen. The Service Module itself was seriously damaged, although the extent of the damage was not evident until the Service Module was jettisoned, shortly before re-entry into Earth's atmosphere, and the crew were able to see and take photographs of the destruction.

The decision was taken to use the Lunar Module, which was fully supplied with oxygen and power, as a 'lifeboat', despite it being designed to support two men for two days, rather than three men for four days. The Lunar Module was dormant for most of the flight, so the decision was taken to 'power up' the Module. Because of the location of the mishap, it was not possible to perform the 'direct abort' path that had been planned for any earlier emergency. It would be necessary for the flight to make a loop around the Moon before heading back to Earth. To do so, changes had to be made to the control systems, so that the engine of the Lunar Module could make the necessary adjustments to the flight path to put the spacecraft on a return trajectory to Earth. These changes were successfully implemented.

Various modifications were made to the Lunar Module to enable it to support the three men for four days. The modifications were successful, despite the crew having to live under extremely uncomfortable conditions, with a lack of power, cold temperatures, high humidity and a shortage of drinking water.

One critical point was to adapt the carbon-dioxide scrubber system to remove excess carbon dioxide from the air. This used lithium hydroxide canisters. Although a certain (very small) percentage of carbon dioxide is required for efficient human respiration, an excess may lead to unconsciousness or even death. There were lithium hydroxide canisters in the Command Module, but they were of the wrong shape to fit the system in the Lunar Module. Luckily, an improvised solution was found, although this involved the use of plastic covers from flight manuals and other items, all held together with duct tape. Fortunately the improvised solution worked, and carbon dioxide levels fell immediately.

The damaged Service Module was jettisoned, with the Lunar Module being undocked somewhat later, before atmospheric re-entry. The Command Module re-entered the atmosphere as planned, and splashed down in the South Pacific on April 17. The three crew members had been safely returned to Earth.

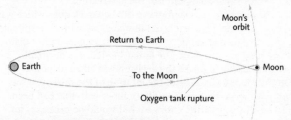

The flight path of the Apollo 13 mission, showing the location where the explosion occurred.

Calendar for April

02		Australian and New Zealand Daylight Saving Time ends
02	08:01	Regulus 4.6°S of the Moon
06	04:34	Full Moon
06	17:23	Spica 3.3°S of the Moon
10	06:25	Antares 1.5°S of the Moon
11	22:10	Mercury at greatest elongation (19.5°E, mag. -0.0)
11	22:27	Jupiter at superior conjunction
13	09:11	Last Quarter
14–30		April Lyrid meteor shower
15–28		π-Puppid meteor shower
16	02:24	Moon at perigee = 367,968 km
16	03:49	Saturn 3.5°N of the Moon
17	17:24	Neptune 2.3°N of the Moon
19–May.28		η-Aquariid meteor shower
19	17:31	Jupiter 0.1°S of the Moon
20	04:12	New Moon
20	04:36	Hybrid solar eclipse
21	07:05	Mercury 1.9°N of the Moon
21	13:00	Uranus 1.7°S of the Moon
22–23		April Lyrid meteor shower maximum
23	07:17	Aldebaran 8.8°S of the Moon
23	13:03	Venus 1.3°S of the Moon
24		π-Puppid meteor shower maximum
26	02:18	Mars 3.2°S of the Moon
26	18:03	Pollux 1.5°N of the Moon
27	21:20	First Quarter
28	06:43	Moon at apogee = 404,299 km
29	16:00	Regulus 4.6°S of the Moon

April 2 • *The Moon is between Regulus and Algieba (as seen from Sydney).*

April 10 • *The waning gibbous Moon is close to Antares, near the southeast (as seen from central USA).*

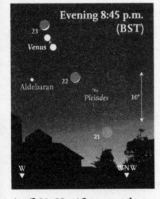

April 21–23 • *After sunset the narrow crescent Moon passes the Pleiades, Aldebaran and Venus (as seen from London).*

April 26 • *The Moon is between Mars and Castor. Pollux and Mars are equally bright: mag. 1.2 (as seen from Sydney).*

CANIS MAJOR
PUPPIS
VELA
Mimosa
CRUX
LUPUS
Eta Car Nebula
60°S
50°S
Omega Centauri
40°S
CENTAURUS
30°S
HYDRA
SCORP
ANTLIA
PYXIS
20°S
CORVUS
LIBRA
MONOCEROS
CRATER
10°S
Alphard
SEXTANS
Spica
HYDRA
VIRGO
0°
EQUATOR
10°N
Regulus
Denebola
LEO
Algieba
20°N
Arcturus
SERP CAP
M44
COMA BERENICES
BOÖTES
CANCER
30°N
CANES VENATICI
Alphecca
LEO MINOR
40°N
60°S
50°N
CORONA BOREALIS
50°S
Pollux
URSA MAJOR
60°N
Castor
MARS
M13
30°S GEMINI
LYNX
HERCULES
20°S
Kochab
Pherkad
10°S
URSA MINOR
0°
North Celestial Pole
DRACO
AURIGA
Capella
Polaris
April Lyrids
10°N
CAMELOPARDALIS
Vega
20°N
Double Cluster
LYRA
30°N
CEPHEUS
Deneb
CYGNUS
PERSEUS
Algol
CASSIOPEIA
40°N
LACERTA
VULPECULA
TAURUS
TRIANGULUM
M31
SAGITTA
Pleiades
50°N
ANDROMEDA
60°N E.CL.
ARIES
PISCES
PEGASUS

northwest   north   northeast

104

April – Looking North

Ursa Major is still high overhead for most northern-hemisphere observers, with **Boötes** and bright **Arcturus** (α Boötis), the star indicated by the 'arc' of the 'tail' of Ursa Major, to its northeast. The small circlet of **Corona Borealis**, with its single bright star of **Alphecca** (α CrB) is still farther to the east. **Cassiopeia** has now swung round, and is almost on the meridian to the north, below **Polaris** and northwest of **Ursa Minor**. **Cepheus** is beginning to climb higher and the head of **Draco** is almost due east of Polaris, but slightly farther south than the two 'Guards' in Ursa Minor, **Kochab** (β UMi) and **Pherkad** (γ UMi). On the other side of the meridian to the northwest lies the constellation of **Auriga**, with bright **Capella**. The inconspicuous constellation of **Camelopardalis** lies between Polaris, Auriga and **Perseus**. The stars of the Milky Way run through Auriga, Perseus, and Cassiopeia towards the much more densely populated regions in **Cygnus** and farther south. Nearly the whole of Cygnus is visible above the horizon in the northeast, where the small constellation of Lyra, with the bright star **Vega**, is clearly seen with **Hercules** above it.

The constellation of Perseus is beginning to descend into the northwest, following the constellation of **Andromeda**, which, for most observers, has now disappeared below the northwestern horizon.

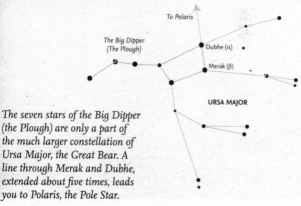

The seven stars of the Big Dipper (the Plough) are only a part of the much larger constellation of Ursa Major, the Great Bear. A line through Merak and Dubhe, extended about five times, leads you to Polaris, the Pole Star.

southeast　　　　　　　　south　　　　　　　　southwest

April – Looking South

The constellation of **Leo** is still conspicuous in the south, although now it is **Denebola** (β Leonis) rather then **Regulus** that is close to the meridian. **Boötes** and **Arcturus** (α Boötis), the brightest star in the northern hemisphere, are high in the southeast. Beyond Leo, along the ecliptic to the east, the zodiacal constellation of **Virgo** is clearly visible, although **Spica** (α Virginis) is the only really bright star in the constellation.

Farther south, **Crux** is now nearly 'upright' and close to the meridian, with **Rigil Kentaurus** and **Hadar** (α and β Centauri, respectively), conspicuous to its southeast. Rigil Kentaurus is a multiple system, with two stars fairly easily visible with a small telescope. A third star in the system, known as **Proxima Centauri**, is much fainter, and is actually the closest star to the Solar System at a distance of about 4.3 light-years. In recent years, it has been found to host the closest known exoplanet, which orbits Proxima every 11.2 days.

The stars of **Vela** and **Carina** lie to the west of Crux and the meridian, while much farther south, and out to the west is **Canopus** (α Carinae) the second-brightest star in the whole sky after Sirius in Canis Major. Even farther south, and right on the horizon for people at 30°S (roughly the latitude of Sydney in Australia) is **Achernar** (α Eridani), the main star in the long, winding constellation of **Eridanus** (the River), which begins near the star Rigel in Orion, far to the north.

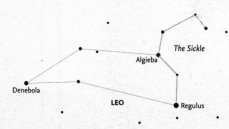

The distinctive constellation of Leo, with Regulus and 'The Sickle' on the west. Algieba (γ Leonis), north of Regulus, appearing double, is a multiple system of four stars.

The Moon Illusion

Frequently, when the Moon is seen rising or setting, and it is close to the horizon, it appears absolutely enormous. It seems far larger than when it is high in the sky. In fact, this is an optical illusion – it is known as the 'Moon Illusion'. Our eyes are playing tricks. In reality, the Moon is exactly the same size wherever it is in the sky. Try covering it with your finger, held at arm's length. Any finger is more than large enough to cover the Moon. (In fact, the Moon is about 30 minutes across, and a finger is about twice that, about 1 whole degree.) There have been many attempts over the years to explain the Moon Illusion, but it seems that the brain automatically compares the size of the Moon with the distant horizon, and assumes that it is at the same distance – whereas it is, of course, much farther away. The first person to correctly explain the effect was the Arab expert on optics, Ḥasan Ibn al-Haytham, known as Alhazen, in a book written in 1011–1021.

When the Moon is seen against distant objects, the brain is tricked into thinking it is much larger than its true size.

May

May – Introduction

May 2023 is rather poor in astronomical events. There is a penumbral lunar eclipse on May 5. Although this is visible from a wide area of Asia and Africa, because it is a penumbral event, there is very little for observers to detect. Even at mid-eclipse none of the Moon will be within the Earth's umbra.

The *η-Aquariid* meteor shower comes to maximum on May 6. This shower, like the Orionids in October, is associated with debris from Comet 1P/Halley. Maximum in 2023 is on Day 16 of the lunation, one day after Full Moon, so conditions are very poor. This, coupled with the fact that the radiant is close to the celestial equator, means that conditions for northern observers are bad, even though the shower does occasionally produce bright meteors that leave persistent trains.

In May 2020, it was announced that two new southern meteor streams had been discovered. These streams are apparently associated with long-period comets. The first of these new meteor streams is the *γ-Piscid Austrinids*, which appear to peak early in the morning of 15 May and may well be an annual shower. This date is on Day 25 of the lunation, four days before New Moon in 2023 so observing conditions are moderately good. The second southern shower is the *σ-Phoenicids* and this appears to peak slightly later on the same date. The data gathered over the preceding years suggest that the rate for this shower may vary considerably from year to year. Both new showers are known from extremely low numbers of meteors only (15 and 14, respectively) and have yet to be recognized as 'official' meteor showers.

On 26 May 2008, NASA's *Phoenix* spacecraft landed in the northern polar region of *Mars*.

The planets

Mercury is mainly lost in twilight throughout May, although it comes to greatest western elongation at the end of the month (on May 29), when it is very close to the horizon. *Venus* is bright (mag. -4.2 to -4.4), and is initially in *Taurus*, and then crosses into *Gemini*. There is an occultation of the planet on May 27, but this is essentially invisible, occurring in daylight and visible only from a small area of the southern Indian Ocean. *Mars* (mag. 1.3 to 1.6) moves from *Gemini* into *Cancer*, in the evening sky. *Jupiter* is inside *Pisces*, lost in the morning twilight. *Saturn* (mag. 1.0) is in *Aquarius*. *Uranus* is in *Aries* at mag. 5.9 and reaches superior conjunction on May 9. *Neptune* remains in *Pisces* at mag. 7.9.

M

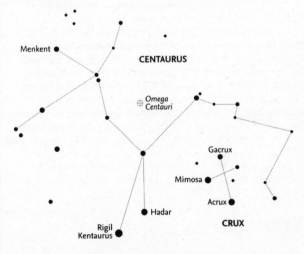

In May, the constellation of Centaurus is well placed for observing from the tropical regions and farther south. It contains the largest and finest globular star cluster in the sky, Omega Centauri. Crux, the Southern Cross is between the front and the hind legs of the Centaur.

Sunrise and sunset

City	Date	Sunrise	Sunset
Buenos Aires, Argentina			
	May 01	10:29	21:11
	May 31	10:51	20:51
Cape Town, South Africa			
	May 01	05:21	16:05
	May 31	05:42	15:45
London, UK			
	May 01	04:33	19:24
	May 31	03:50	20:08
Los Angeles, USA			
	May 01	13:04	02:36
	May 31	12:43	02:58
Nairobi, Kenya			
	May 01	03:28	15:32
	May 31	03:29	15:32
Sydney, Australia			
	May 01	20:30	07:15
	May 31	20:51	06:55
Tokyo, Japan			
	May 01	19:49	09:27
	May 31	19:27	09:50
Washington, DC, USA			
	May 01	10:11	24:00
	May 31	09:45	00:26
Wellington, New Zealand			
	May 01	19:08	05:28
	May 31	19:37	05:01

NB: the times given are in Universal Time (UT)

The Moon's phases and ages

Northern hemisphere

Full Moon

| 01 | 02 | 03 | 04 | 05 17:34 | 06 | 07 | 08 |
| Day 11 | Day 12 | Day 13 | Day 14 | Day 15 | Day 16 | Day 17 | Day 18 |

Last Quarter

| 09 | 10 | 11 | 12 14:28 | 13 | 14 | 15 | 16 |
| Day 19 | Day 20 | Day 21 | Day 22 | Day 23 | Day 24 | Day 25 | Day 26 |

New Moon

| 17 | 18 | 19 15:53 | 20 | 21 | 22 | 23 | 24 |
| Day 27 | Day 28 | Day 29 | Day 0 | Day 1 | Day 2 | Day 3 | Day 4 |

First Quarter

| 25 | 26 | 27 15:22 | 28 | 29 | 30 | 31 | |
| Day 5 | Day 6 | Day 7 | Day 8 | Day 9 | Day 10 | Day 11 | |

Southern hemisphere

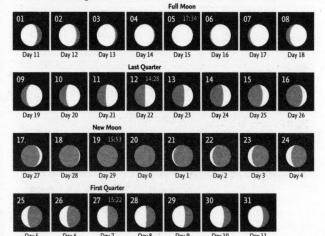

Full Moon

| 01 | 02 | 03 | 04 | 05 17:34 | 06 | 07 | 08 |
| Day 11 | Day 12 | Day 13 | Day 14 | Day 15 | Day 16 | Day 17 | Day 18 |

Last Quarter

| 09 | 10 | 11 | 12 14:28 | 13 | 14 | 15 | 16 |
| Day 19 | Day 20 | Day 21 | Day 22 | Day 23 | Day 24 | Day 25 | Day 26 |

New Moon

| 17 | 18 | 19 15:53 | 20 | 21 | 22 | 23 | 24 |
| Day 27 | Day 28 | Day 29 | Day 0 | Day 1 | Day 2 | Day 3 | Day 4 |

First Quarter

| 25 | 26 | 27 15:22 | 28 | 29 | 30 | 31 | |
| Day 5 | Day 6 | Day 7 | Day 8 | Day 9 | Day 10 | Day 11 | |

M

The Moon

On May 4, one day before Full Moon, the Moon is 3.3° north of *Spica* (mag. 0.98) in *Virgo*. On May 5 (that is, at Full Moon) there is a penumbral lunar eclipse. Invisible to the naked eye, mid-eclipse is over the Indian Ocean. Two days later, the Moon is 1.5° north of *Antares*. By May 13, one day after Last Quarter, the Moon is 3.3° south of *Saturn* (mag. 1.0) in *Aquarius* in the morning sky. It passes 0.8° north of *Jupiter* in *Pisces* on May 17, with the planet lost in the morning twilight. On May 19, at New Moon, it is 1.8° north of *Uranus* (mag. 5.9) in *Aries*. The Moon is 8.7° north of *Aldebaran* on May 20 and 2.2° north of *Venus* on May 23. It passes 1.6° south of *Pollux* on May 24, and, later the same day, is 3.8° north of *Mars*. The Moon is 4.6° north of *Regulus* in *Leo* on May 27. It is again 3.3° north of *Spica* on May 31.

Antikythera Mechanism
On 17 May 1902, divers recovered a remarkable mechanism from the wreck of a Roman vessel that had foundered off the small island of Antikythera in the Aegean Sea. In 1905, a philologist studying the markings on the device first suggested that it was used for astronomical calculation. It has since become known as the Antikythera Mechanism.

The highly-corroded device was initially taken to the National Archaeological Museum in Athens. There were no conservation techniques that would prevent it from splitting into no less than 82 fragments. It is not impossible that other fragments remain on the seabed.

Highly sophisticated techniques, such as modern computer-assisted 3-D X-ray tomography, have now become available for study of the device. These have revealed not only some of the complex internal gearing but also some of the written instructions on the outside of the device. It is now obvious that the mechanism was an extremely complicated (and amazingly accurate) device for the computation of astronomical events, in particular the motions of the Sun, Moon and planets, and for predicting forthcoming eclipses. The date at which the device was created remains uncertain, lying within the range of about 200 BCE to around 60 BCE. There is scientific disagreement about the starting date at which the mechanism was set, with dates of 178 and 204 BCE being proposed. (The latest computation favours the earlier date.)

The device has proved to be extremely complex, integrating results using the Metonic cycle of lunar phases and eclipses as well as the Callippic cycle that matches the tropical (solar) year and the monthly lunar calendar. Both of these cycles are approximately correct, although they differ from the true values by slight amounts. They could, however, be used for calculation purposes over a reasonably short period of time (measured in tens of years, rather than many centuries). Additional gearing provided details of the positions of the various planets (from Mercury to Saturn), the Sun and the Moon, along the ecliptic. Dials provided information on the lunar and solar calendars, the timing of the Pan-Hellenic Games (the Olympiads) and the Saros cycle for predicting lunar and solar eclipses.

Although there have been various technical descriptions of studies of the device, published in specialized scientific journals, more recent 'popular' accounts include:

'Wonder of the Ancient World', *Scientific American*, January 2022, pages 24 to 33

'Ancient computer may have had its clock set to 23 December 178 BC', *New Scientist*, 16 April 2022, page 9

The main fragment of the Antikythera Mechanism, showing the four spokes of the main drive wheel (left) and the rear of the fragment (right).

Calendar for May

01	23:28	Mercury at inferior conjunction
04	01:33	Spica 3.3°S of the Moon
05	17:22	Penumbral lunar eclipse
05	17:34	Full Moon
06		η-Aquariid meteor shower maximum
07	13:10	Antares 1.5° S of the Moon
09	19:56	Uranus at superior conjunction
11	05:05	Moon at perigee = 369,343 km
12	14:28	Last Quarter
13	13:07	Saturn 3.3°N of the Moon
15	01:25	Neptune 2.2°N of the Moon
17	13:18	Jupiter 0.8°S of the Moon
18	01:35	Mercury 3.6°S of the Moon
19	00:22	Uranus 1.8°S of the Moon
19	15:53	New Moon
20	15:55	Aldebaran 8.7°S of the Moon
23	12:08	Venus 2.2°S of the Moon
24	02:14	Pollux 1.6°N of the Moon
24	17:32	Mars 3.8°S of the Moon
26	01:39	Moon at apogee =404,509 km
27	00:09	Regulus 4.6°S of the Moon
27	15:22	First Quarter
29	05:34	Mercury at greatest elongation (24.9°W, mag. 0.5)
31	10:44	Spica 3.3°S of the Moon

May 3 • The Moon is in the company of Spica. There are no other bright stars nearby (as seen from central USA).

May 7–8 • The Moon passes Antares, in the south-southeast. Sabik is nearby (as seen from London).

M

May 22–24 • The waxing crescent Moon passes Venus, Pollux, Castor and Mars (as seen from central USA).

May 26 • The Moon is almost First Quarter, when it passes between Regulus and Algieba (as seen from central USA).

May – Looking North

The constellation of **Cassiopeia** is now low over the northern horizon, to the east of the meridian, to observers at mid-northern latitudes. To its west, the southern portions of both **Perseus** and **Auriga** are becoming difficult to observe as they become closer to the horizon, although the **Double Cluster**, between Perseus and Cassiopeia, is still clearly visible.

High above, **Ursa Major** has started to swing round to the west and the stars that form the extended portion towards the south (the 'legs' and 'paws' of the 'bear') are becoming easier to see. **Alkaid** (η Ursae Majoris), at the end of the 'tail', is almost exactly at the zenith for observers at 50° North. The whole of the constellations of **Cepheus**,

---

On 17 May 1902, the Antikythera Mechanism was recovered from a wreck (see pages 114–115).

Draco, **Lyra** and **Ursa Minor** are easy to see, together with the inconspicuous constellation of **Camelopardalis**. For those people with the keenest eyesight, they can even make out the long, straggling line of faint stars forming the constellation of **Lynx** in the northwestern sky, beyond the extended portion of Ursa Major. These faint constellations become difficult to detect during the brighter nights towards the end of the month.

Also high in the sky for northern observers are the constellations of **Boötes**, with bright **Arcturus**, and **Hercules**, with its clearly visible 'Keystone' of four stars, on one side of which is **M13**, the finest globular cluster in the northern sky. The small, but striking arc of stars forming the constellation of **Corona Borealis** lies between Hercules and Boötes. It has just a single bright star, **Alphecca** (α Coronae Borealis).

Much of **Cygnus** is clearly seen in the east, as are two (**Deneb** and **Vega**) of the three stars that form the Summer Triangle. The third star, **Altair** in Aquila, begins to climb above the horizon late in the night and later in the month. Across the Milky Way, between Cassiopeia and Cygnus, and below Cepheus, is the tiny zig-zag of stars that forms the small, and often ignored, constellation of **Lacerta**.

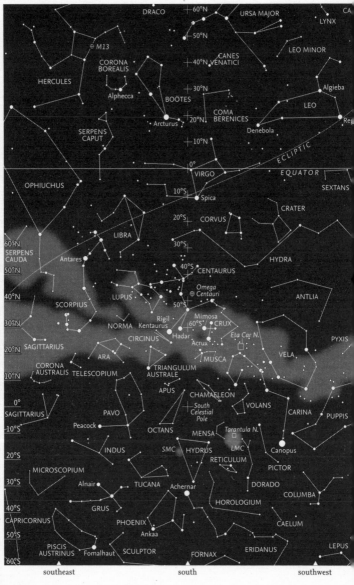

May – Looking South

Spica (α Virginis) in the sprawling constellation of **Virgo** is now close to the meridian. The constellation of **Boötes**, with brilliant **Arcturus** (the brightest star in the northern hemisphere of the sky) is higher above and slightly to the east, clearly visible to observers in the north. Following Virgo across the sky is the rather undistinguished constellation of **Libra** and, following it, and rising in the east, **Scorpius**, with the red supergiant star **Antares** and the distinctive line of stars, running south and ending in the 'sting'. (As Orion sinks in the west, so Scorpius rises in the east. This recalls one of the many legends about Orion: with his being pursued by a scorpion, sent to distract him while fighting.) Above Libra and Scorpius, the large constellation of **Ophiuchus** is beginning to climb higher in the sky. It lies between the two portions of **Serpens** (the only constellation to be divided into two parts). **Serpens Caput** (the Head of the Serpent) lies west of Ophiuchus, between it and Boötes, and **Serpens Cauda** (the Tail of the Serpent) is to the east, between Ophiuchus and the Milky Way.

Farther south, between Scorpius and the two brightest stars in **Centaurus** (**Rigil Kentaurus** and **Hadar**), lies the constellation of **Lupus**, just east of the meridian and lying along the Milky Way. **Crux** is now more-or-less 'upright', with the small constellation of **Musca** below it. Below the bright pair of stars in Centaurus (α and β Centauri) is the constellation of **Triangulum Australe**, a much larger and more striking constellation than its counterpart (Triangulum) in the north. Lying between Rigil Kentaurus and Triangulum Australe is the very tiny and indistinct constellation of **Circinus**.

The stars of **Vela** and **Carina** are becoming lower in the southwest, following the constellation of **Puppis** and brilliant **Canopus** (α Carinae) down towards the horizon. Puppis and the **Large Magellanic Cloud** (LMC) are close to the horizon for anyone at the latitude of 30°S (about the latitude of Sydney in Australia), where **Achernar** (α Eridani) is now too low to be seen.

There are several, small, relatively faint constellations in this part of the sky. The most distinct is probably **Pavo**, with its single bright star (α Pavonis), known as **Peacock**. Other constellations are **Apus**, **Chamaeleon**, **Octans** (which actually includes the South Celestial Pole), **Mensa** and **Volans**.

Eclipses

Solar eclipses

Between two and five solar eclipses may occur in any one year.
During the period when dates have been reckoned on the basis
of the Gregorian calendar (initially introduced in 1582), there
have been six years in which there have been five solar eclipses.
The last was 1935 and the next one will be 2206. There were
four partial solar eclipses in 2011, but there will be no partial
eclipses in 2023.

There are three (or four) types of solar eclipse. In a total
eclipse, the whole of the Sun's disc is covered. This will
normally occur when the Moon is close to perigee (closest
to the Earth) and thus appears to subtend the greatest angle
(diameter) from Earth. Such a situation occurs along a very
narrow, central eclipse track across the surface of the Earth. The
width of this track depends entirely upon the exact distance of
the Moon from the Earth. When the Moon is farther from the
Earth, particularly when it is at apogee (farthest away) it may not
be large enough to cover the whole of the Sun's disc. We then
have an annular eclipse, when, from certain locations a ring (an
annulus) of the Sun's disc remains visible. Again, such annular
eclipses are visible from a very restricted track on the surface
of the Earth. Outside the central eclipse tracks, whether total or
annular, only part of the Sun's disc is covered, giving a partial
eclipse. A very rare form of eclipse is known as a hybrid eclipse

*The annular eclipse on 4 January 1992 as photographed from California
as the Sun was setting over the Pacific Ocean.*

(or an annular/total eclipse). In this, the eclipse is total for a very short duration in the central portion of its track and annular at the beginning and end of the event. This occurs because (on the spherical Earth) the two ends of the track are at different (greater) distances from the Moon than the central portion. The solar eclipse of 20 April 2023 will be a hybrid event.

The magnitude of any eclipse is the fraction of the Sun's disc that is covered by the Moon. For the two partial solar eclipses of 2022 (April 30 and October 25) the magnitudes were approximately 0.64 and 0.86, respectively (0.6389 and 0.8611, to be precise). Obviously, for a total eclipse, the magnitude will be 1 or greater. In annular eclipses, the magnitudes are obviously less than 1, and in the rare hybrid eclipses, the magnitude changes from being less than 1 at the beginning and end to slightly greater than 1 during the central phase.

Authorities disagree wildly on how much of the Sun must be covered for any change in illumination to be detected by normal human eyesight. The values quoted lie between 20 and 90 per cent. (Some of this wide range is due to differences in what is being discussed. In some cases it is the actual level of illumination, in others it is the percentage of the Sun's disc that has to be covered.) Certainly at least 70 per cent of the Sun needs to be covered for any alteration in the amount of sunlight to be readily detected. (A similar consideration applies to penumbral lunar eclipses, which are not detected by the human eye, and which only become visible when the Moon passes partially or completely within the umbra – the darkest part of the Earth's shadow.) If viewed with appropriate methods, partial solar eclipses are visible over fairly large areas of the Earth. The partial eclipse of 30 April 2022 was visible over a wide region of the southeastern Pacific, off the western coast of South America, and that of 25 October 2022 was over a very large area of Asia, northeastern Africa, and most of Europe.

Lunar eclipses

In contrast to total and annular solar eclipses, total lunar eclipses are visible from a wide area of the Earth – in effect from anywhere the Moon is above the horizon. In any lunar eclipse, the Moon, as it moves eastward against the background stars, first enters the outer region of the Earth's shadow (the penumbra), where part of the Sun's disc is covered by the Earth. As it moves farther east, it encounters the region (the umbra) where the whole disc is hidden by the Earth. After the central phase, it will, of course, pass into the penumbra, and finally leave the shadow entirely. As mentioned, the diminution of light from the Moon is generally not detectable by the human eye until the Moon enters the umbra. The exact classification of an eclipse will depend on the proportion of the Moon that enters the umbra. The eclipse of 5 May 2023 is entirely penumbral, none of the Moon enters the Earth's umbra. That of 28 October 2023 is partial, as a very tiny portion of the Moon does enter the umbra.

In many cases the Moon's path does not take it fully within the umbra, giving a partial eclipse. Technically, even if a small section of the Moon remains within the penumbra, most being darkened within the umbra, the eclipse is still described as a partial eclipse. (The eclipse of 19 November 2021 was of this sort.) Both lunar eclipses in 2022 (of May 16 and November 8) were total, where the whole of the Moon passed into the central umbra. During the total phases, the colour of the Moon changes markedly. It is generally reddened, because red light penetrates through the ring (the annulus) of the Earth's atmosphere that an observer on the Moon would see around the Earth. The exact shade depends greatly upon the amount of cloud cover and may also be affected if there have been volcanic eruptions on Earth that have ejected material to great altitudes, where it tends to block the light, causing a dark eclipse. On very rare occasions, especially when it has passed through the centre of the shadow, the Moon has darkened to such an extent that it has even been said to have effectively disappeared. Frequently, the portion of the Moon that is covered by the edge of the shadow, i.e., close to the penumbra, appears a bluish shade, where some shorter-wavelength light passes through the Earth's outer atmosphere. Generally, some of the brighter features of the Moon, such as the craters Aristarchus, Proclus and Copernicus, remain detectable, even during the central phases of an eclipse.

June

June – Introduction

June is a very quiet month for astronomical activity. There are no major meteor showers and because of the light nights, normal astronomy is difficult for northern observers. Time to turn one's attention to 'daytime astronomy' – the study of the Sun.

There is a commonly-held view that the weather in June is hot because the Earth is then closest to the Sun. But this is completely wrong. The Earth's orbit is not a true circle, but an ellipse. In 2023 the distance varies from 147,098,855.22 km (0.983295578 AU) on January 4 (at perihelion) to 152,093,178.67 km (1.016680584 AU) on July 6 (at aphelion). So the Earth is actually closest to the Sun in January. The winter solstice (for northern-hemisphere residents) when the Sun is lowest in the sky, is on December 22 in 2023 (December 21 in 2022). At the summer solstice (winter for those in the southern hemisphere) on June 21, it is actually farther from the Sun, reaching its farthest point on July 6.

The cause of the seasons, with warm summers and cold winters, is solely caused by the tilt of the Earth's axis to the plane of its orbit around the Sun – the ecliptic. As such, the northern hemisphere is most strongly tilted towards the Sun in June and July, and the southern hemisphere in November and December. At the equinoxes (March 20 and September 23 in 2023), the Earth's axis is tilted so that the planet is 'side on' to the Sun, so day and night are of (approximately) equal length and the Sun's heat is evenly distributed between the two hemispheres. (See pages 80 and 81 for a description of the equinoxes.)

On 27 June 2018 the Japanese spaceprobe *Hayabusa 2* reached the minor planet *Ryugu*.

There is an almost exact cycle in the Sun's magnetic orientation – it switches north and south every 11 years. The cycle is clearly displayed by the number of sunspots and active areas that are found on the surface of the Sun, although these are actually signs of the underlying magnetic activity. So every 11 years, the cycle varies from solar minimum (very few sunspots and active areas) to maximum (many large sunspots, active areas and flares) and back to minimum again. At present we are entering Cycle 25, and there has been a sharp rise in activity since solar minimum. The peak of Cycle 25 is predicted to occur in 2025, but predicting the Sun's activity is fraught with problems, because the processes within the Sun are poorly understood.

The planets

Mercury is low in the morning twilight. It may be possible to glimpse it early in the month, when it is farthest from the Sun, but rapidly becomes invisible. *Venus*, moving steadily eastwards, brightens from mag. -4.4 to -4.7 over the month. It is at eastern elongation on June 4, when its magnitude is 4.4. *Mars* is in *Cancer* at mag. 1.6 to 1.7 and moves into *Leo* at the end of the month. *Jupiter* is in the morning sky in *Aries* at mag. -2.1 to -2.2. *Saturn* (mag. 0.9 to 0.8) is moving slowly in *Aquarius*. *Uranus* remains in *Aries* at mag. 5.8 and *Neptune* is in *Pisces* at mag. 7.9.

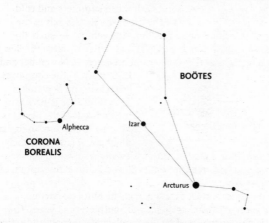

In June, the constellations of Boötes (with Arcturus) and Corona Borealis (with Alphecca) are well placed for observation. Arcturus has an orange tint and is the brightest star in the northern celestial hemisphere, at magnitude -0.05.

Sunrise and sunset

City	Date	Sunrise	Sunset
Buenos Aires, Argentina			
	Jun. 01	10:52	20:51
	Jun. 30	11:01	20:53
Cape Town, South Africa			
	Jun. 01	05:43	15:45
	Jun. 30	05:52	15:47
London, UK			
	Jun. 01	03:49	20:09
	Jun. 30	03:47	20:22
Los Angeles, USA			
	Jun. 01	12:43	02:59
	Jun. 30	12:45	03:09
Nairobi, Kenya			
	Jun. 01	03:29	15:32
	Jun. 30	03:35	15:38
Sydney, Australia			
	Jun. 01	20:52	06:54
	Jun. 30	21:01	06:57
Tokyo, Japan			
	Jun. 01	19:26	09:51
	Jun. 30	19:29	10:01
Washington, DC, USA			
	Jun. 01	09:45	00:27
	Jun. 30	09:46	00:38
Wellington, New Zealand			
	Jun. 01	19:38	05:01
	Jun. 30	19:48	05:01

NB: the times given are in Universal Time (UT)

The Moon's phases and ages

Northern hemisphere

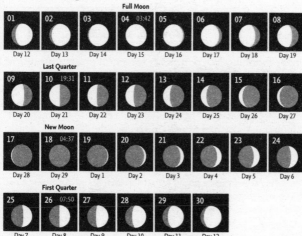

			Full Moon				
01	02	03	04 03:42	05	06	07	08
Day 12	Day 13	Day 14	Day 15	Day 16	Day 17	Day 18	Day 19

Last Quarter							
09	10 19:31	11	12	13	14	15	16
Day 20	Day 21	Day 22	Day 23	Day 24	Day 25	Day 26	Day 27

	New Moon						
17	18 04:37	19	20	21	22	23	24
Day 28	Day 29	Day 1	Day 2	Day 3	Day 4	Day 5	Day 6

First Quarter					
25	26 07:50	27	28	29	30
Day 7	Day 8	Day 9	Day 10	Day 11	Day 12

Southern hemisphere

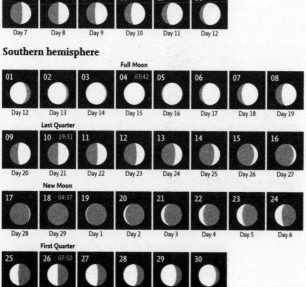

			Full Moon				
01	02	03	04 03:42	05	06	07	08
Day 12	Day 13	Day 14	Day 15	Day 16	Day 17	Day 18	Day 19

Last Quarter							
09	10 19:31	11	12	13	14	15	16
Day 20	Day 21	Day 22	Day 23	Day 24	Day 25	Day 26	Day 27

New Moon							
17	18 04:37	19	20	21	22	23	24
Day 28	Day 29	Day 1	Day 2	Day 3	Day 4	Day 5	Day 6

First Quarter					
25	26 07:50	27	28	29	30
Day 7	Day 8	Day 9	Day 10	Day 11	Day 12

J

The Moon

On June 3, one day before New Moon, the Moon is 1.5° north of *Antares* in *Scorpius*. By June 9, waning gibbous and one day before Last Quarter, it is 2.0° south of *Saturn* (mag. 0.9) in *Aquarius*. By June 11, it is 2.0° south of *Neptune* (mag. 7.9) in *Pisces*. On June 14, the Moon is 1.5° north of *Jupiter*, in *Aries*, low in the morning sky. The next day, June 15, it passes 2.0° north of *Uranus* (mag. 5.8) also in Aries. On June 20, the waxing crescent Moon is 1.7° south of *Pollux* in *Gemini*. By June 22, the Moon is 3.7° north of brilliant *Venus* (mag. -4.6) in *Cancer*. Later the same day it passes 3.8° north of *Mars*. The next day, June 23, the Moon is 4.4° north of *Regulus* in *Leo*. By June 27, it is 3.1° north of *Spica* in *Virgo*.

Strawberry Moon

June has always been noted for strawberries, and this appears in the name of the Full Moon this month (on June 14). This name was used on both sides of the Atlantic. The Choctaw tribe of southeastern America had different names for Full Moons that occurred in early or late June. In early June it was 'Moon of the peach' and in late June it was 'Moon of the crane'. Other names are 'Hot Moon' and 'Rose Moon' and in the Old World, 'Mead Moon'.

Transits of Venus

Transits of Venus across the face of the Sun are actually quite rare. They occur in pairs with an interval of about 8 years. The last was in 2012 (preceded by the one in 2004) and the next will be in 2117. Transits are rare because of the requirements that the orbits of the Earth and Venus should be in line, and also that the orbital periods should combine. Obviously, the timing and the location of the two planets and their orbits need to be in the correct alignment for a transit to occur. This recurs in a pattern of some 243 years, with long intervals of over 121 and 105 years between pairs of transits.

The Persian astronomer Avicenna (980 to 1037 CE) claimed to have seen Venus on the face of the Sun. There was a transit in 1032 CE, but his observation is disputed and he may have mistaken a sunspot for the planet.

The first to predict a transit was Johannes Kepler, who, in 1627, predicted the transit of 1631. But this was not observed. The first confirmed observations were of the transit of 4 December 1639 by Jeremiah Horrocks and William Crabtree.

In 1691, the famous astronomer Edmond Halley predicted a transit of Venus in 1761 and suggested that this could be used to refine the scale of the Solar System, and primarily determine the value of the Astronomical Unit (the distance between the Earth and the Sun) by making observations from widely separated points on the surface of the Earth. Although Halley died in 1742, his predictions were respected and numerous expeditions were mounted, some of which were successful. The second transit of the pair, in 1769, was also observed from many different locations, although timings were complicated by what has come to be known as the 'black drop effect', a dark bar that appeared at the edge of the planet's disc at the various contacts. This was long thought to be caused by the thick atmosphere of Venus, but has since been shown to be an optical illusion crested by turbulence in the Earth's own atmosphere. In 1771, the French astronomer Jérôme Lalande used the observations of 1761 and 1769 to calculate the Astronomical Unit's value as 153 million km. (This may be compared with the modern value of 149,597,870 km.)

Transits also occurred in 1874 and 1882, and the latest pair in 2004 and 2012.

The transit of Venus on 8 June 2004 as photographed from Degrania A, in Israel.

Calendar for June

03	21:53	Antares 1.5°S of the Moon
04	03:42	Full Moon
04	05:00 *	Uranus (mag. 5.8) 2.9°N of Mercury (mag. 0.1)
04	11:01	Venus at greatest elongation (45.4°E, mag. -4.4)
06	23:06	Moon at perigee = 364,861 km
09	20:23	Saturn 3.0°N of the Moon
10	19:31	Last Quarter
11	07:46	Neptune 2.0°N of the Moon
14	06:35	Jupiter 1.5°S of the Moon
15	09:54	Uranus 2.0°S of the Moon
16	20:38	Mercury 4.3°S of the Moon
16	23:09	Aldebaran 8.7°S of the Moon
18	04:37	New Moon
20	09:47	Pollux 1.7°N of the Moon
21	14:58	June solstice
22	00:48	Venus 3.7°S of the Moon
22	10:09	Mars 3.8°S of the Moon
22	18:30	Moon at apogee = 405,385 km
23	07:46	Regulus 4.4°S of the Moon
26	07:50	First Quarter
27	19:47	Spica 3.1°S of the Moon

These objects are close together for an extended period around this time.

June 3 · *The almost Full Moon is close to Antares, with Sabik a little farther to the southeast (as seen from London).*

June 14 · *Low in the east, Jupiter is close to the Moon. Hamal is about ten degrees higher (as seen from London).*

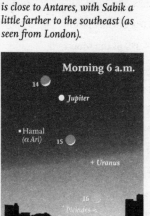

June 14–16 · *In the early morning, the Moon passes Jupiter, Uranus (mag. 5.8) and the Pleiades (as seen from Sydney).*

June 20–23 · *The Moon passes Pollux, Venus and Mars. Then passes between Regulus and Algieba (as seen from Sydney).*

J

northwest   north   northeast

June – Looking North

For observers in the northern hemisphere, the time around the summer solstice (June 21) is frustrating for observing, because a form of twilight persists throughout the night. The sky remains so light that most faint stars and constellations are invisible unless conditions, such as the lack of light pollution, are particularly favourable. Even the highly distinctive set of seven stars forming the asterism of the 'Plough' (or 'Big Dipper') in **Ursa Major** may be difficult to see. The constellation has now swung round to the west, but the fainter stars of the constellation, lying even farther to the west, are harder to distinguish. The constellation of **Boötes** with bright **Arcturus** (α Boötis) is high overhead for northern observers, and best seen when facing south. The sprawling constellation of **Hercules** lies between **Lyra** and Boötes.

The four stars forming the 'head' of **Draco** lie east of the meridian, although only the brightest, **Eltanin** (γ Draconis) and, possibly **Rastaban** (β Draconis) are clearly visible. Farther north, in **Ursa Minor**, are **Polaris** itself and the two 'Guards', **Kochab** (β UMi) and **Pherkad** (γ UMi).

For observers around 50°N, the southern portions of the constellation of **Auriga** are partially lost on the northern horizon, although bright **Capella** (α Aurigae) should still be visible. Much of the neighbouring, fainter constellation of **Perseus** is difficult to make out. The two main stars of **Gemini**, **Castor** (α Geminorum) and **Pollux** (β Geminorum), the star closest to the ecliptic, and occasionally occulted by the Moon, are low on the northwestern horizon. Somewhat higher in the sky, and northeast of the meridian is **Cassiopeia** with **Cepheus** above it.

Two of the stars forming the angles of the distinctive Summer Triangle, **Deneb** (α Cygni) in **Cygnus** and **Vega** (α Lyrae) in Lyra are clearly visible as, for much of the night, is the third star, Altair (α Aquilae) in **Aquila**. Deneb lies in the Milky Way, at the beginning of the dark Great Rift that runs down the constellation and where obscuring dust prevents us from seeing the dense star clouds of the Milky Way itself.

June – Looking South

The inconspicuous constellation of **Libra** is on the meridian, but the sky is really dominated by the striking constellation of **Scorpius**, slightly to the east. These two were once a single constellation, of course, until the 'claws' of the scorpion were formed into the constellation of The Balance. Although it is rather low for most northern observers, to those farther south, fiery red **Antares** (α Scorpii) is a major beacon and the whole constellation has a very significant presence in the winter skies. The constellation of **Virgo**, and bright **Spica** is now well to the west, along the ecliptic.

At this time of the year, the next constellation in the zodiac, **Sagittarius**, is becoming clearly seen, having risen in the east. The main body of the constellation forms the asterism known as 'The Teapot', but the constellation also has a long, curving chain of faint stars to the south, rather similar to the long line of stars below Scorpius, although curving in the opposite direction. This chain of stars partially encloses the small constellation of **Corona Australis**, although this constellation is not as well-formed as its northern counterpart (Corona Borealis). Rising in the east and following Sagittarius into the sky is the constellation of **Capricornus**. To the west of the 'sting' of Scorpius is the rather untidy tangle of stars that forms the constellation of **Lupus** and part of **Centaurus**.

The small constellation of **Crux** is now well south of the two brightest stars of Centaurus, **Rigil Kentaurus** and **Hadar**. The constellations of **Vela** and **Carina** and the **'False Cross'** are plunging towards the horizon. For observers at the latitude of Sydney in Australia, both **Canopus** (α Carinae) and **Achernar** (α Eridani) are so low that they are often invisible through absorption at the horizon.

The South Celestial Pole in **Octans** is surrounded by faint constellations: Octans itself, **Chamaeleon**, **Volans**, **Mensa**, **Hydrus**, **Tucana**, and **Indus**. Only **Pavo**, to the northeast and **Grus**, to the southeast, are slightly more distinct. However, this area also includes the **Large Magellanic Cloud** (LMC), the largest satellite galaxy to our own. Most of the LMC lies in the constellation of **Dorado**, although some extends into the neighbouring constellation of Mensa.

Observing the Sun

Observing the Sun requires special equipment and great care. Never be tempted to view the Sun directly through any telescope or binoculars, even when it is low in the sky, at sunrise or sunset. Although the Sun then appears red, with all the other colour being scattered away, and also much dimmer, its infra-red radiation still penetrates the atmosphere, and if this is concentrated onto your eye's retina by a telescope or binoculars, it can produce blindness. So, don't do it!

There are, of course, specialized telescopes, designed for observing the Sun, but these, containing what are known as Hα filters, are expensive. 'Eclipse glasses', which consist of thin metal foil, are safe, as are very dark welders' glasses, but the glass 'solar filters' sold with some cheap telescopes are actually dangerous. They may break when subjected to heating from the Sun and suddenly allow damaging radiation through to the eye. Once again – don't use them! It has sometimes been said that eclipses may be viewed through smoked glass or exposed photographic film, but both of these are unsafe.

There are ways of viewing the Sun directly, but these require the use of full-aperture specialized solar filters. They may be of metal-coated Mylar film – which tend to give the Sun a blue colour – or special, metal-coated glass. The latter are better, but far more expensive (and fragile). Any such filters need to cover the complete aperture of any telescope.

The simplest way of viewing the Sun and seeing sunspots is probably by projection. Set up a refracting telescope or binoculars on a tripod, pointing at the Sun. (Reflecting telescopes may be used, but they are generally less convenient.) Catadioptric telescopes (those that incorporate both mirrors and lenses) are usually difficult to use and normally need to be reduced in aperture – otherwise heat from the Sun can cause damage to the telescope itself.

But be careful if the telescope has a finder attached. Use an utterly opaque cover to fit over the main objective lens of the finder. Similarly, block off one side of the binoculars if you are using those. Hold a piece of paper, or preferably, a card, behind the eyepiece. A circular image of the Sun will be projected onto the card. (You can use the image to ensure that you are pointing the telescope in the right direction.) Move the card towards and away from the eyepiece, until you get a sharp image, on which you should see any sunspots that are on that side of the Sun.

If you plan to do a lot of solar observing, you may wish to make up a simple, lightweight, open-sided 'box' to hold your card or paper at the correct distance from the eyepiece. Because the distance may be varied easily, still with a sharp image, observers often make the size to fit the standard blanks supplied by many observing organisations. A frame of balsa wood is usually sufficient, with card for the sides. This will be light in weight and so may be fitted easily to the telescope, and not cause great problems over the balance. If the 'box' allows you to use a standard, paper observing blank, you can draw any sunspots directly onto the paper. Provided the box shields the interior from most stray light, the image of the Sun will be relatively easily seen in considerable detail.

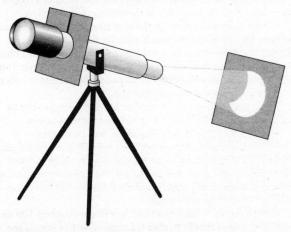

A simple method of observing the Sun. Point a telescope (or binoculars), mounted on a tripod, at the Sun – making sure any finder (or one half of the binoculars) is completely covered. Features on the Sun (or, in this case, an eclipse) may be seen by holding a card behind the eyepiece.

Sunspots usually consist of a black centre (the umbra) surrounded by a lighter, grey area (the penumbra). Close to the limb (the edge) of the Sun you may be able to see bright areas (known as faculae) and close to the centre of the image some of the solar granulation may be detected. Granules are actually the tops of convection cells within the Sun's surface. There may also be tiny black dots (known as 'pores') which are incipient sunspots. These appear in the gaps between individual granules.

Sunspots, which are actually areas where the surface is very slightly cooler than their surroundings, often occur in pairs, where there is a loop in the magnetic field. Sunspots frequently arise where the ends of the magnetic-field loop leave and then re-enter the Sun's surface.

Very, very rarely, when observing the sun (by projection), apart from dark sunspots, you may happen to be looking at the image when there is a temporary, brilliant flash of light. This is a solar flare. It actually occurs when the Sun's invisible magnetic field lines reconnect in such a way that there is a great release of energy. Such solar flares may release pulses of particles into space, which cross the immense distance between the Sun and Earth, where they may cause aurorae, geomagnetic storms, or radio blackouts – or all three effects. Only a few flares have ever been seen by visual observers, so do not expect to ever see one.

On 30 June 1908 an extremely violent airburst occurred in the Podkamennaya Tunguska (Stony Tunguska) River region of Siberia. The explosion flattened an estimated 80 million trees over an area of about 2,150 km^2. There is controversy over whether the impacting body was asteroidal or cometary in nature. It was most probably an airburst at an altitude of between 5 and 10 kilometres, caused by a stony meteoroid some 50–60 m in diameter.

July

July – Introduction

Northern observers will continue to suffer from light nights during July, although there still remains the chance that they may be able to see noctilucent clouds during the first half of the month. (The noctilucent-cloud 'season' in the north lasts about 6 to 8 weeks, centred on the summer solstice.)

The Earth reaches aphelion, the farthest point from the Sun in its yearly orbit, at 20:07 UT on July 6. Its distance from the Sun is then 1.016680584 AU (which equals 152,093,178.67 km).

The first meteor shower that is active in July is that of the *α-Capricornids,* visible from both hemispheres. This is a very long shower, beginning in early July and continuing until about mid-August. Unfortunately, the maximum rate is very low, around 5 meteors per hour, although there are occasional bright meteors. The peak occurs on July 30. Both at the beginning of the shower and at maximum the Moon is nearly Full, so conditions are very unfavourable. The parent body is a comet known as 169P/NEAT, one of the few comets discovered by NASA's Near-Earth Asteroid Tracking (NEAT) programme.

The second shower active in July is the *Southern δ-Aquariids,* also visible from both hemispheres. These begin in mid-month (on July 12, two days after Last Quarter) and continue until August 23. Their maximum, like the α-Capricornids, is on July 30. Their hourly rate is higher than the α-Capricornids, and is about 25 meteors per hour. The parent body is uncertain, but believed to be Comet 96P/Macholz.

There is a minor southern shower, the *Piscis Austrinids,* which is a moderately long shower, beginning about July 15 and continuing until August 10. Like the α-Capricornids, the maximum hourly rate is very low, around 5 meteors per hour, with maximum on July 28, a few days before Full Moon. The parent body is currently unknown.

The most important shower that begins in July is the *Perseids.* These begin very favourably on July 17, at New Moon, but there is a strong maximum on August 12–13, so are more appropriately described next month.

The planets

Mercury reaches superior conjunction on July 1, but is too close to the Sun to be readily visible, although it may be glimpsed in the evening sky towards the end of the month. **Venus** is very bright (mag. -4.7 to -4.4), being brightest (mag. -4.7) until mid-month, so may be visible although close to the Sun. **Mars** is mag. 1.7 to 1.8 in **Leo**. **Jupiter** is mag. -2.3 in **Aries**. Saturn is in **Aquarius**, beginning its retrograde motion early in the month towards opposition on August 27. **Uranus** at mag. 5.8 is in **Aries** and is on the border of **Taurus** at the end of the month. **Neptune** remains in **Pisces** at mag. 7.9.

The Japanese spacecraft **Hayabusa 2** obtained sub-surface samples from the asteroid **Ryugu** on 11 July 2019.

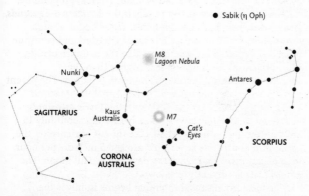

The sky looking south is dominated by the two zodiacal constellations of Scorpius and Sagittarius, on either side of the meridian, although they are low for observers at most northern latitudes.

Sunrise and sunset

City	Date	Sunrise	Sunset
Buenos Aires, Argentina			
	Jul. 01	11:01	20:53
	Jul. 31	10:48	21:12
Cape Town, South Africa			
	Jul. 01	05:52	15:48
	Jul. 31	05:40	16:06
London, UK			
	Jul. 01	03:47	20:22
	Jul. 31	04:22	19:52
Los Angeles, USA			
	Jul. 01	12:45	03:09
	Jul. 31	13:04	02:56
Nairobi, Kenya			
	Jul. 01	03:35	15:38
	Jul. 31	03:37	15:41
Sydney, Australia			
	Jul. 01	21:01	06:57
	Jul. 31	20:48	07:15
Tokyo, Japan			
	Jul. 01	19:29	10:01
	Jul. 31	19:48	09:47
Washington, DC, USA			
	Jul. 01	09:47	00:38
	Jul. 31	10:08	00:22
Wellington, New Zealand			
	Jul. 01	19:48	05:02
	Jul. 31	19:30	05:25

NB: *the times given are in Universal Time (UT)*

The Moon's phases and ages

Northern hemisphere

Full Moon

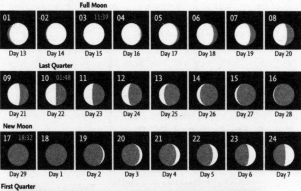

Last Quarter

New Moon

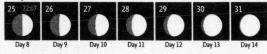

First Quarter

Southern hemisphere

Full Moon

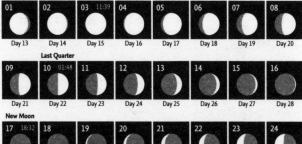

Last Quarter

New Moon

First Quarter

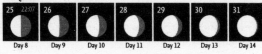

J

145

The Moon

The Moon is 1.5° north of *Antares* in *Scorpius* on July 1, two days before Full Moon. On July 7, four days after Full, it is 2.7° south of *Saturn* (mag. 0.8), which is retrograding in *Aquarius*. By July 11, the Moon is 2.2° north of *Jupiter* (mag. -2.3) in *Aries*. On July 12 the waning crescent is 2.3° north of *Uranus* (mag. 5.8). By July 14 it is 8.8° north of *Aldebaran*. Three days later, at New Moon, it is 1.8° south of *Pollux* in *Gemini*. On July 19 it passes 3.5° north of Mercury (mag. -0.5), low in the evening twilight. The next day the waxing crescent is 7.9° north of brilliant *Venus* (mag. -4.6), also low in the evening sky. Later that day, the Moon is 4.2° north of *Regulus*, between it and *Algieba* (γ Leonis). On July 21 the waxing crescent is 3.3° north of *Mars*, also in *Leo*. Before First Quarter on July 25 it is 2.8° north of *Spica* in *Virgo*. By July 28 it is 1.2° north of *Antares*, very slightly closer to the star than at the beginning of the month.

Buck Moon

One of the names used in North America for the Full Moon for the month of July is 'Buck Moon'. This term derives from the fact that this is when new antlers grow on the heads of male (buck) deer. There are many other names for this Full Moon: among the Chippewa and Ojibwe of the Great Lakes area, following from the 'Strawberry Moon' of June, it is the 'Raspberry Moon'. Then, among the Arapaho of the Great Plains and the Omaha of the Central Plains it is 'the Moon when the buffalo bellow'. There are yet other terms from the Old English/Anglo-Saxon calendar, including the 'Thunder Moon', 'Wort Moon', and 'Hay Moon'

Venus

The planet Venus is at its brightest (mag. -4.7) during July. It is the only planet that is so bright that it may cast shadows, and is even visible during the daytime. (Because it is generally close to the Sun, great care must be taken when searching for it, although at elongations in the evening or morning sky, it may be clearly visible after the Sun has set, or before it has risen.) The planet's brightness is largely caused by its dense cloud cover, which completely hides the surface.

Venus has been visited by many different spaceprobes and it has frequently been used to give gravity-assists to spacecraft journeying to other planets. (The first of these was by **Mariner 10** in 1974, on the probe's journey to Mercury.) Because of the extremely hostile environment with exceptionally high surface pressures (approximately 93 times that on Earth) and temperatures (over 400°C) with a highly corrosive atmospheric layer of suphuric acid, only a few landers have made it to the surface (primarily the Soviet **Venera** series).

Venus was the first planet to be the object of any spaceprobe's mission (**Venera 1** in 1961), the first to be successfully visited (by **Mariner 2** in 1962) and the first to have a surface lander (**Venera 7** in 1970). Because the extremely dense atmosphere and heavy clouds shield the surface from direct observation, the atmosphere itself has been extensively studied.

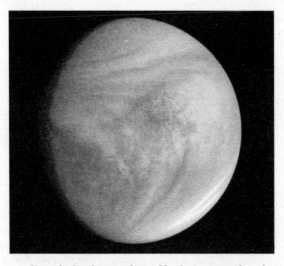

An ultraviolet-band image obtained by the Japanese Akatsuki spacecraft on 23 December 2016, showing the complex structure of the clouds on Venus.

J

Ultraviolet photographs of the clouds provided information about their structure and the extremely high winds (~300 km/hr) in the upper atmosphere.

Information was obtained about the composition and structure of the atmosphere both during fly-by missions, during the descent of the *Venera* landers and also from the NASA *Pioneer Multiprobe* descent module in 1978. Examination of the atmosphere has continued, most notably with ESA's *Venus Express* mission, which orbited the planet from 2006 to 2014, and carried out extensive atmospheric observations. The European investigation of the atmosphere will be continued by the proposed *Envision* mission. Currently, the Japanese spacecraft *Akatsuki* is in orbit around Venus, carrying out atmospheric observations.

Surface observations have proved more difficult, because of the dense cloud layers, and have largely come from radar observations. The first were results from the Arecibo telescope in the 1970s. These were followed in 1978 by observations by

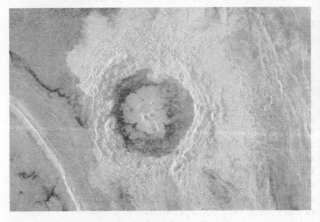

The crater Dickinson on Venus. The fact that the ejecta are largely confined to the right-hand side suggests that the crater was formed by an oblique impact, with the body travelling from the left.

the NASA *Pioneer Venus* orbiter, which enabled the first true topographic map of the surface to be constructed.

Some details of part of the surface were provided by the later *Venera* missions. Far more significant were radar observations from the NASA *Magellan* probe, which was placed into low orbit in 1994. This provided global mapping of the surface, and gave rise to the view that the surface was largely volcanic, but that any activity (particularly plate-tectonic activity) had ceased long ago, leaving the surface appearing geologically young, even though its age is actually some 500 million years. The lack of craters is taken as an indication of the young age of the surface, although the thick atmosphere tends to protect the surface from all but the most massive impactors. This concept of a lack of geological activity has been challenged in recent years, especially with the discovery of short-term changes in the atmospheric composition, which may indicate current volcanic activity.

The surface consists of remarkably uniform volcanic plains, with few distinctive features. The highest region is Maxwell Montes (about 11 km high), where the temperature is approximately 380°C, and the pressure about 45 times that on the surface of the Earth. Both reductions from those generally prevailing on the surface are caused by the height of the feature.

In mid-2021 and early 2022, it was confirmed that at least three missions to Venus are under development: *VERITAS* (Venus Emissivity, Radio Science, InSAR, Topography, and Spectroscopy), a NASA orbiter to be launched in 2027 or 2028, *DAVINCI+* (Deep Atmosphere Venus Investigation of Noble gases, Chemistry, and Imaging) an orbiter and atmospheric probe, and *EnVision*, an ESA orbiter mission. These missions are partly inspired by the fact that more and more Venus-like exoplanets are being discovered, and also by a desire to study the geological history of a planet that is close to the Earth in size, but differs so greatly in current conditions. The lack of any apparent plate-tectonic activity may be an indication of the lack of water in the early history of Venus. Examination of the way in which Venus has undergone extreme global warming from the greenhouse effect of its carbon-dioxide atmosphere is yet another reason for it to be studied in detail.

J

Calendar for July

01	05:06	Mercury at superior conjunction
01	07:55	Antares 1.5°S of the Moon
03–Aug.15		α-Capricornid meteor shower
03	11:39	Full Moon
04	22:25	Moon at perigee = 360,149 km
06	20:07	Earth at aphelion (1.016680584 AU)
07	03:10	Saturn 2.7°N of the Moon
07	19:58	Minor planet (15) Eunomia at opposition (mag. 8.7)
08	14:12	Neptune 1.7°N of the Moon
10	01:48	Last Quarter
11	21:21	Jupiter 2.2°S of the Moon
12–Aug.23		Southern δ-Aquariid meteor shower
12	17:48	Uranus 2.3°S of the Moon
14	05:05	Aldebaran 8.8°S of the Moon
15–Aug.10		Piscis Austrinid meteor shower
17–Aug.24		Perseid meteor shower
17	16:22	Pollux 1.8°N of the Moon
17	18:32	New Moon
19	08:56	Mercury 3.5°S of the Moon
20	06:57	Moon at apogee = 406,289 km
20	08:37	Venus 7.9°S of the Moon
20	14:32	Regulus 4.2°S of the Moon
21	04:00	Mars 3.3°S of the Moon
25	03:41	Spica 2.8°S of the Moon
25	22:07	First Quarter
26	13:00 *	Mercury (mag. -0.2) 5.3°N of Venus (mag. -4.5)
28		Piscis Austrinid meteor shower maximum
28	17:46	Antares 1.2°S of the Moon
30		α-Capricornid meteor shower maximum
30		Southern δ-Aquariid meteor shower maximum

These objects are close together for an extended period around this time.

July 7 • *The Moon is close to Saturn. Fomalhaut is much closer to the horizon (as seen from London).*

July 12–13 • *In the northeast the Moon passes Jupiter, Uranus (mag. 5.8) and approaches the Pleiades (as seen from Sydney).*

July 19–20 • *The Moon is in the company of Venus and Mars, when it passes between Regulus and Algieba (as seen from central USA).*

July 29 • *Three planets join Regulus: Venus, Mercury and Mars (as seen from Sydney).*

J

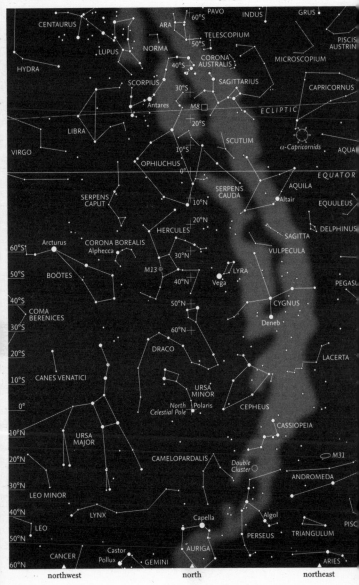

northwest   north   northeast

July – Looking North

The brilliant star **Vega** (α Lyrae) is now shining high overhead and the constellations of **Hercules** and **Lyra** are on opposite sides of the meridian, not far from the zenith for observers at 40°N, while it is the head of **Draco** that is near the zenith for observers slightly farther north at 50°N. Beyond Hercules, in the northwestern sky, are the constellations of **Corona Borealis** and **Boötes**, the latter with brilliant, orange-tinted **Arcturus**.

The stars of the Milky Way are now running more-or-less 'vertically', from north to south on the eastern side of the meridian. The constellation of **Cygnus** is 'upside down', high in the sky. For observers at the equator, it is the giant constellation of **Ophiuchus** that is at the zenith, with the two parts of **Serpens** (**Serpens Caput** to the west, and **Serpens Cauda** to the east, among the clouds of the Milky Way). The third star of the Summer Triangle, **Altair** (α Aquilae) in **Aquila** is similarly high in the sky.

Ursa Major is now clearly visible in the northwest, and on the opposite side of the meridian, the constellation of **Cepheus**, with its base in the Milky Way, is at a slightly greater altitude. **Cassiopeia**, the other constellation that, like Ursa Major, is the key to to finding one's way around the northern circumpolar constellations, lies in the Milky Way on the opposite side of the North Celestial

NASA's **Viking 1** spacecraft was the first successful Martian lander, touching down in western Chryse Planitia on 20 July 1976. It operated until 11 November 1982.

Pole and **Polaris** in **Ursa Minor**. The faint constellations of **Camelopardalis** and **Lynx** lie to the west and slightly farther south. The chain of faint stars forming Lynx runs 'horizontally' below the outflung stars of Ursa Major. Below Cassiopeia on the other side of the meridian, **Perseus**, with the famous variable star, **Algol**, is beginning to climb higher in the sky and observers at mid-northern latitudes will find that they can now more clearly see **Capella** (α Aurigae) and the northernmost portion of **Auriga**. Observers in the far north (around latitude 60°N) may even occasionally glimpse **Castor** and **Pollux** in **Gemini** peeping above the northern horizon.

July – Looking South

The sky looking south is dominated by the two zodiacal constellations of **Scorpius** and **Sagittarius**, which lie on the ecliptic on either side of the meridian, although they are low for observers at most northern latitudes. Bright, red **Antares** (α Scorpii) is very conspicuous, even when it is low in the sky. Not for nothing has it earned the name that means 'Rival of Mars'. The roughly triangular shape of **Capricornus**, the next zodiacal constellation, lies east of Sagittarius and is now clearly visible. Below Scorpius, in the Milky Way, are the small, and often ignored, constellations of **Norma** (which is little more than three stars and easily overlooked) and **Ara**, the latter with a more distinctive shape and brighter stars. The stars of **Lupus** and the outlying stars of **Centaurus** (including the great globular cluster known as **Omega Centauri**) lie farther west.

High above, the constellation of **Ophiuchus** lies at the zenith for observers on the equator, with **Aquila** and bright **Altair** (α Aquilae) to the east. Northwest of Ophiuchus is **Boötes**, the principal star of which is **Arcturus**, which most people see as having an orange tint. Between Arcturus and the meridian is the circlet of stars that forms the constellation of **Corona Borealis**.

South of Scorpius is **Triangulum Australe**, to the east of the two bright stars of Centaurus, **Rigil Kentaurus** and **Hadar**, and lying within the star clouds of the Milky Way. At about the same altitude is the constellation of **Pavo** and, beyond it, **Indus**. Still farther east lies the elongated and somewhat distorted cross-shape that is the constellation of **Grus**. **Piscis Austrinus**, which lies south of Capricornus and Aquarius, has a single bright star, **Fomalhaut** (α Piscis Austrini). This ancient constellation (it was mentioned by Ptolemy, in the second century AD) has now risen in the east.

Crux and the adjoining constellation of **Musca** are now low on the horizon for observers at the equator, and the **Small Magellanic Cloud** (SMC) in **Hydrus** is actually below it. For observers farther south, **Achernar** (α Eridani) is now clearly visible, as is the constellation of **Phoenix** to the east. Only observers in the extreme south, however, will be able to see the whole of **Vela** and **Carina** as well as brilliant **Canopus** (α Carinae).

J

Following its successful return of samples from minor planet *(162173) Ryugu* in December 2020, spaceprobe *Hayabusa 2* has been targetted to fly by the body known as *(98943) 2001 CC21* in July 2026 and rendezvous with another, *(1998) KY26* in July 2031.

August

August – Introduction

Meteors

August sees one of the finest meteor showers of the year. These are the famous **Perseids**, which are well-known, even to the general public, partly because they are visible during the warm nights of summer. In some Catholic countries they are known as the 'Tears of Saint Lawrence' because they are visible on August 10, the date of his martyrdom. They are said to represent sparks from the fire on which he was burnt. The Perseids are a very long shower, generally beginning about July 17 and continuing until around August 24, with a maximum on August 12–13, when the rate may reach as high as 100 meteors per hour (and on rare occasions, even higher). In 2023, maximum is when the Moon is a waning crescent, three days after Last Quarter, so conditions are reasonably good. The radiant lies in the northern portion of Perseus, towards the border with Camelopardus.

The Perseids are debris from Comet 109P/ Swift-Tuttle (the Great Comet of 1862) and there is a minor concentration that may give an early minor peak the day before the nominal maximum. Perseid meteors are fast and many of the brighter ones leave persistent trains. Some of the meteors may also produce bright fireballs.

There are three other meteor showers, all of which reach their peak

On 18 August 1976 the Soviet lunar probe **Luna 24** landed on the Mare Crisium. After obtaining a surface sample, it lifted off from the Moon the following day, and returned the sample to Earth on 22 August 1976.

maxima in July, but still exhibit activity that persists into August. These are the **α-Capricornids**, which may be active until 14 August (just before New Moon), the **Southern δ-Aquariids** (lasting until August 24) and a southern shower, the **Piscis Austrinids**, which may persist until August 27. All these three showers have been described more fully on pages 142–143.

One short, minor, northern meteor shower, the **α-Aurigids**, begins late in the month, on August 28 in 2023, just three days before Full Moon, and reaches its weak peak, with 5 or 6 meteors per hour, on September 1.

The planets

Mercury is low in the morning twilight. On August 10 it reaches greatest elongation east when it is 27.4° from the Sun, at mag. 0.3. **Venus** is initially mag. -4.3, but rapidly fades to mag. -4.1, then brightens to mag. -4.6). Because of its magnitude at the end of the month, it may be detectable in the morning twilight. **Mars** (mag. 1.8) may just be visible at the beginning of evening twilight, but is a difficult object. **Jupiter** is clearly visible in **Aries** at mag. -2.4 to -2.6. **Saturn**, in **Aquarius**, is mag 0.4 when it reaches opposition on August 27. **Uranus** remains in **Aries** at mag. 5.9 and **Neptune** is in **Pisces** at mag. 7.8.

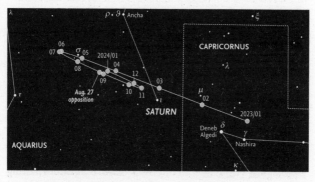

The path of Saturn in 2023. It starts in the constellation of Capricornus, then crosses into Aquarius halfway through February and comes to opposition on August 27 (see also page 172).

Sunrise and sunset

City	Date	Sunrise	Sunset
Buenos Aires, Argentina			
	Aug. 01	10:47	21:13
	Aug. 31	10:14	21:34
Cape Town, South Africa			
	Aug. 01	05:39	16:06
	Aug. 31	05:06	16:27
London, UK			
	Aug. 01	04:24	19:50
	Aug. 31	05:11	18:50
Los Angeles, USA			
	Aug. 01	13:04	02:55
	Aug. 31	13:26	02:22
Nairobi, Kenya			
	Aug. 01	03:37	15:41
	Aug. 31	03:31	15:36
Sydney, Australia			
	Aug. 01	20:47	07:15
	Aug. 31	20:14	07:36
Tokyo, Japan			
	Aug. 01	19:49	09:46
	Aug. 31	20:12	09:11
Washington, DC, USA			
	Aug. 01	10:09	00:21
	Aug. 31	10:36	23:40
Wellington, New Zealand			
	Aug. 01	19:28	05:26
	Aug. 31	18:47	05:55

NB: *the times given are in Universal Time (UT)*

The Moon's phases and ages

Northern hemisphere

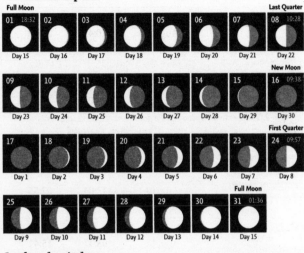

Southern hemisphere

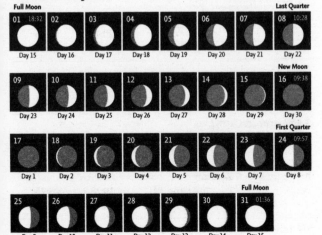

The Moon

On August 3, just after Full Moon, it is 2.5° south of **Saturn** in **Aquarius**. By August 8, at Last Quarter, it is 2.9° north of **Jupiter** in **Aries**. The next day, on August 9, it is 2.6° north of **Uranus** (mag. 5.8), close to **Taurus**. The next day, August 10, it is 9.0° north of **Aldebaran**. On August 13, a waning crescent, it is 1.7° south of **Pollux** in **Gemini**. On August 15, it is 13.3° north of **Venus** (mag. -4.1) in **Pisces** in the early evening sky. At New Moon on August 16, it is 4.1° north of **Regulus**. On August 18 it is 2.2° north of **Mars** on the edge of evening twilight. On August 21 it is 2.5° north of **Spica** in **Virgo**. By August 25, one day after First Quarter, it is 1.1° north of **Antares** in **Scorpius**. By August 30, one day before Full Moon, it is again 2.5° south of **Saturn** in **Aquarius** as it was at the beginning of the month.

Sturgeon Moon

The Algonquin tribes of North America called the Full Moon of August the 'Sturgeon Moon' because of the numerous examples that they captured in the lakes where they fished. To the Cree of the Canadian Northern Plains it was the 'Moon when young ducks begin to fly'. Many tribal names refer to the berries ripening at this time: 'berries', 'black cherries', and 'chokeberries', are just a few examples. Others refer to the fact that corn is ripening, such as among the Ponca of the Southern Plains where it was the 'Corn is in the Silk Moon'. To the Haida of Alaska it was the 'Moon for cedar bark for hats and baskets'. In the ancient Old English/Anglo-Saxon calendar it was sometimes known as 'Barley Moon', 'Fruit Moon' or 'Grain Moon'.

Sagittarius and Scorpius

Although they are low on the horizon for most observers in the northern hemisphere, these two constellations contain much of interest. Sagittarius, in particular, has many fascinating objects. It contains the **Teapot** asterism, which is extremely distinctive. Because the constellation harbours the centre of the Milky Way galaxy, it has no fewer than eleven Messier objects and many additional globular clusters and emission nebulae. Other objects are undoubtedly hidden by the vast amounts of interstellar dust that lie in this region. There are a few 'windows' in this dust, however, that allow more distant objects to be seen. One such object is **NGC 6822**, 'Barnard's Galaxy', an irregular galaxy in the far northwest of the constellation, which lies outside our own stellar system and is part of the Local Group of galaxies. Much closer to the centre of the galaxy is the area known as 'Baade's Window' where there is little obscuring dust, enabling two distant globular clusters to be seen.

Almost due north of the tip of the Teapot's 'Spout' (the star **Alnasl** or γ **Sgr**), is the galactic centre, and the location of the (invisible) supermassive black hole known as 'Sagittarius A*' (abbreviated as **Sgr A***). The mass of this black hole has been determined from the orbits of several stars around the centre. The current estimate is that it probably contains more than 4 million solar masses. One star, known as S4714, has been determined to be the star that comes closest to the central black hole (at 'perinigricon') at a distance of about 12.6 AU, roughly as close as Saturn comes to the Sun. (Just as we have 'perigee' for

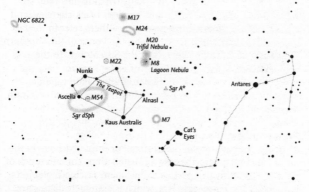

The constellation of Sagittarius not only contains the galactic centre (Sgr A*) but also many interesting objects. The satellite galaxy, the Sagittarius Dwarf Spheroidal Galaxy (Sgr dSph), is hidden by clouds of dust.

A

A highly detailed image of the Lagoon Nebula (M8) recently obtained by the VLT Survey Telescope (VST) at Paranal Observatory in Chile.

the closest approach to Earth, and 'perihelion' for the closest approach to the Sun, we now have 'perinigricon' for the closest approach to a black hole.)

Slightly southwest of the Galactic Centre is **M8**, the 'Lagoon Nebula', which is just visible to the naked eye under good conditions. This is a giant star-forming interstellar cloud, primarily a giant emission nebula, illuminated by the young, hot stars within it. Slightly farther north again is **M20**, the 'Trifid Nebula', which is actually a rare combination of three different forms of nebulosity: an emission nebula (pinkish), where the radiation from young stars is causing the surrounding hydrogen gas to glow; a reflection nebula (bluish in tint); and a dark dust nebula that is causing the dark 'lanes', dividing the object into three lobes. Very slightly to the north of M20 and farther east is **M22**, a bright (mag. 5) globular cluster, visible to the naked eye. One of Messier's objects, **M24**, northeast of Kaus Borealis (λ Sgr) is not a true open cluster, but consists of a dense concentration of stars, known as the Small Sagittarius Star Cloud. (The Large

Sagittarius Star Cloud is the dense portion of the galaxy's central bulge, visible below – southwest of – the Dark Rift.)

Much farther north, and near the point that divides the constellations of Serpens and Scutum from Sagittarius, is **M17**, which is known both as the 'Omega Nebula' and also as the 'Swan Nebula'. This is a gaseous nebula, best seen in binoculars or a telescope.

The globular cluster **M54** lies southeast of the star *Ascella* (ζ Sgr) that is the base of the 'Handle' of the 'Teapot'. It is actually part of the Sagittarius Dwarf Spheroidal Galaxy (*Sgr dSph*) which was discovered much later (1999) than M54 itself, and which is now known to be one of the Galaxy's satellite galaxies.

Just over the southwestern border of the constellation and into the constellation of Scorpius, slightly farther south than the star *Kaus Australis* (ε Sgr), the base of the 'Spout' of the 'Teapot', is **M7** (sometimes known as the Ptolemy Cluster), a very large, bright open cluster, that is mag. 3.5 and thus readily visible to the naked eye, but best seen with low magnification in binoculars.

An all-sky image of the Galaxy and the two Magellanic Clouds obtained by the Gaia spacecraft. The satellite galaxy, the Sagittarius Dwarf Spheroidal Galaxy (Sgr dSph), is a barely-visible streak below the galactic centre.

A

Calendar for August

01	18:32	Full Moon
02	05:52	Moon at perigee = 357,311 km
03	10:26	Saturn 2.5°N of the Moon
04	22:02	Neptune 1.5°N of the Moon
08	09:44	Jupiter 2.9°S of the Moon
08	10:28	Last Quarter
09	01:03	Uranus 2.6°S of the Moon
10	01:47	Mercury at greatest elongation (27.4°E, mag. 0.3)
10	10:43	Aldebaran 9.0°S of the Moon
12–13		Perseid meteor shower maximum
13	11:15	Venus at inferior conjunction
13	22:14	Pollux 1.7°N of the Moon
15	17:00	Venus 13.3°S of the Moon
16	09:38	New Moon
16	11:54	Moon at apogee = 406,635 km (most distant of year)
16	20:37	Regulus 4.1°S of the Moon
18	11:26	Mercury 6.9°S of the Moon
18	23:07	Mars 2.2°S of the Moon
21	10:08	Spica 2.5°S of the Moon
24	09:57	First Quarter
25	02:06	Antares 1.1°S of the Moon
27	07:39	Minor planet (8) Flora at opposition (mag. 8.3)
27	08:28	Saturn at opposition (mag. 0.4)
28–Sep.05		α-Aurigid meteor shower
30	15:54	Moon at perigee = 357,181 km
30	18:08	Saturn 2.5°N of the Moon
31	01:36	Full Moon

August 3–4 • *The Moon passes Saturn, in the western sky. On August 3, it is close to Deneb Algedi (as seen from Sydney).*

August 8–9 • *The Moon passes Jupiter, Uranus (mag. 5.8) and the Pleiades. Aldebaran is nearby (as seen from London).*

August 13 • *The crescent Moon forms an almost isosceles triangle with Pollux and Castor (as seen from central USA).*

August 24 • *The Moon is close to Antares. Larawag is about ten degrees closer to the horizon (as seen from central USA).*

A

August – Looking North

The Milky Way is now running up the eastern side of the sky, where **Cassiopeia** may be found within it, high in the northeast. The constellation of **Cepheus**, with the 'base' of the constellation, which is shaped like the gable-end of a house, lies within the edge of the Milky Way, and the whole constellation is 'upside-down' slightly farther north than Cassiopeia. For observers south of the equator, these constellations are close to the horizon and **Ursa Minor** is on the horizon.

For most mid-latitude northern observers, **Perseus** is now clearly visible, as is the northern portion of **Auriga** with brilliant **Capella** (α Aurigae). Perseus straddles a narrow portion of the Milky Way and although not particularly distinct, a significant section of the Milky Way actually runs through Auriga in the direction of **Gemini**. Auriga contains numerous open clusters and a few emission nebulae, but is without globular clusters. Beyond Perseus in the east lies **Andromeda** and the small constellation of **Triangulum**. Slightly farther north than Perseus and on the other side of the meridian is **Ursa Major**, and for most observers even the stars in the southern, extended portion are visible.

For observers at about 40°N, both **Lyra** and **Cygnus**, with their brilliant principal stars, **Vega** and **Deneb**, respectively, are high overhead, near the zenith. The third star marking the other apex of the Summer Triangle, **Altair** in **Aquila**, is much farther to the south, and all three stars are best seen when looking south. The constellation of **Hercules** lies farther west of Lyra and Cygnus, with most of the constellation of **Pegasus** to the east. The tiny, but highly distinctive constellation of **Delphinus** lies between the Milky Way and the outlying stars of Pegasus. Again, these areas are best seen when looking south.

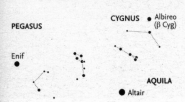

There are four small constellations between Cygnus, Pegasus, and Aquila. They are, from left to right: Equuleus, Delphinus, Sagitta, and Vulpecula.

August – Looking South

The star clouds of the Milky Way dominate the sky at this time of the year. The three stars forming the Summer Triangle, **Deneb** (α Cygni), **Vega** (α Lyrae) and **Altair** (α Aquilae) are clearly visible, as is the Great Rift, running south from near Deneb. The dark Great Rift marks the location of dense clouds of dust that obscure the light from the many stars in the main plane of the Galaxy. Both **Cygnus** and **Aquila** represent birds, and in modern charts these are shown flying 'down' the Milky Way, towards the south. Older charts tended to show Aquila as flying 'across' the Milky Way, in the direction of **Aquarius**, in the east.

For observers at northern latitudes, the constellations of **Hercules** and **Pegasus** are visible, one on each side of the meridian, together with **Delphinus** on the east. Between Cygnus and Aquila, in the Milky Way, lie the two small constellations of **Vulpecula** and **Sagitta**.

The two zodiacal constellations of **Scorpius** and **Sagittarius** are still clearly visible, although becoming rather low for observers at mid-northern latitudes, with even the bright red supergiant star of **Antares** in Scorpius becoming difficult to see. (For observers at 50°N, it is skimming the horizon.) North of Scorpius is the large constellation of **Ophiuchus**, with part of **Serpens**, **Serpens Cauda**, lying in front of the Great Rift.

Capricornus is clearly seen, as is **Aquarius**, the next constellation along the ecliptic. For observers farther south, the curl of stars forming **Corona Australis** lies south of Sagittarius and, farther east, **Piscis Austrinus** with its single bright star, **Fomalhaut** (α Piscis Austrini) is at about the same altitude, with the chain of faint stars that curves south from Sagittarius and the faint constellation of **Microscopium** in between the two. South of Piscis Austrinus is the constellation of **Grus**, with the undistinguished constellation of **Indus** between it and **Pavo**, which is on the meridian.

Farther south, **Lupus**, **Centaurus** and **Crux** are descending towards the horizon, while for observers at 30°S, **Vela** has disappeared, as has most of **Carina**, including brilliant **Canopus**.

A

(8) Flora

The minor planet (8) Flora was discovered in October 1847 by J.R. Hind, and the name was proposed by John Herschel, William Herschel's son. It orbits in the main asteroid belt between Mars and Jupiter and is the largest asteroid orbiting close to the Sun. It's brightness may reach mag. 7.9 at favourable oppositions (it will be mag. 8.3 on August 27 in 2023). It is believed that it was originally a differentiated body, but suffered a cataclysmic collision that created numerous fragments. This impact has been dated by the analysis of cosmic-ray exposure to about 468 million years ago.

Flora is thought to be an aggregation of some of the fragments, rather than a solid body. Its spectrum indicates that it consists of both silicate rocks and nickel-iron metallic portions.

Some of the fragments have formed the Flora family of asteroids. These are believed, on good evidence, to be the source of what are known as the 'L chondrite' meteorites, which amount to about 35 per cent of all those so far catalogued. In the name, the 'L' indicates that these meteorites are low in metals, in contrast to the 'H chondrites'. The L chondrites contain between 4 and 10 per cent of iron-nickel metal,

A finder chart for the minor planet (8) Flora around opposition on August 27. It is then only magnitude 8.3. Background stars are shown down to magnitude 8.5. This is the same day as Saturn is at opposition (see page 159).

An iron meteorite from the fall (known as 'the Sikhote-Alin Iron Rain') that was seen on 12 February 1947. It shows the characteristic 'thumbprints', known as regmaglypts, on its surface.

however, so are magnetic. There are significant indications that the parent body was subject to strong heating before it was fragmented.

The origin of meteorites

Some meteorites may be linked to particular sources. This is particularly the case when meteorites are recovered after known falls. In many instances this enables the orbit of the body before it encountered Earth to be determined, thus leading to its origin.

The most common type is a stony meteorite, like the one recovered after the Chelyabinsk fireball of 15 February 2013. Again, the majority of meteorites are chondrites, containing small, approximately spherical, silicate particles that are thought to have formed in the molten state when free-floating in space. Some chondrules may have remained in this free-floating state, whereas others have been incorporated into a parent body.

A small number of meteorites (only about 6 per cent) consist of a nickel-iron mixture. These tend to survive passage through the atmosphere, whereas the stony type tend to break up on deceleration. The largest known meteorite, the Hoba West meteorite, found in Namibia, is an iron meteorite and is estimated to weigh some 60 tonnes.

Falls of iron meteorites may be very large. On 12 February 1947 a fireball was observed over the Sikhote-Alin mountains in eastern Siberia. About 23 tonnes of materials were recovered.

A

Some meteorites are a mix of stony material and metal. These are (not surprisingly) classed as stony-iron meteorites. They may contain stony particles or crystals, held in a metallic matrix. The Esquel meteorite (illustrated) is of this type. The actual origin of such meteorites is disputed. They may arise from material from the boundary of the mantle and core of a differentiated body, or else from a mixture of materials from a core and mantle that have been created by an impact.

Some meteorites are linked to a particular parent body. This is the case for the meteorite (the Motopi Pan meteorite) that was observed to fall in Botswana on 2 June 2018. This is a form known as a Howardite-Eucrite-Diogenite (HED), which may have arisen from an impact with the minor planet Vesta or one of the Vesta family of asteroids. However, the origin of the HED meteorites from Vesta is disputed.

A few young meteorites have been found to have a composition similar to that of the specimens returned from the Moon by the Apollo and Luna missions. Similarly, another group of young meteorites appears to have come from Mars. (Meteorites may be dated by the amount of exposure to cosmic rays that they have experienced. The majority will have spent a very long time – millions or even billions of years – in space.)

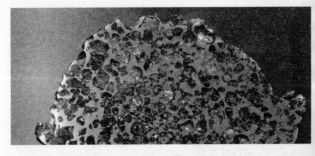

A cut and polished slice of the Esquel meteorite, classed as a stony-iron pallasite. (Esquel is a town in Argentinian Patagonia.) Large olivine crystals (actually yellow-green in colour) are encased in an iron-nickel matrix. This specimen is held in the Royal Ontario Museum.

September

September – Introduction

The (northern) autumnal equinox occurs on September 23 in 2023, when the Sun moves south of the equator in the western side of the constellation of Virgo. This position is sometimes known as the *First Point of Libra*, which is where it lay in former times. However, precession has now shifted this point over the border and into the constellation of *Virgo* (see the map below).

Meteors

After the major Perseid shower in August, there is very little meteor shower activity in September. One minor northern shower, called the *α-Aurigids*, tends to have two peaks of activity. The principal peak occurs on September 1. In 2023, the Moon is one day past Full Moon, so conditions are very poor, apart from the fact that the maximum rate is only 5 or 6 meteors per hour, although the meteors are bright and relatively easy to photograph. Activity from this shower may even extend into October. The *Southern Taurid* shower begins this month (on September 10) and, although rates are similarly low, it often produces very bright fireballs. This is a very long shower, lasting until about November 20, with a maximum in 2023 on October 10–11. This year, activity begins when the moon is a waning crescent (Day 25), and peaks at about the same age (about Day 26 or 27 of the lunation). As a slight compensation for the lack of shower activity, however, the number of sporadic meteors visible in the month of September reaches its highest rate at any time during the year.

The constellation of Virgo, showing the location of the First Point of Libra.

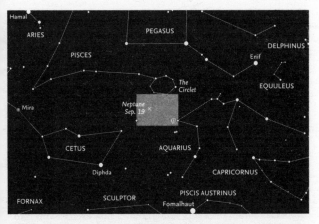

A finder chart for the position of Neptune at its opposition on September 19. The grey area is shown in more detail below.

The path of Neptune in 2023. Stars down to magnitude 8.5 are shown.

The Planets

Mercury is generally lost in morning twilight, but comes to greatest western elongation (17.9°) at mag. -0.5 on September 22. *Venus* is in *Cancer*, visible in the morning sky and very bright at mag. -4.7. *Mars*, in *Virgo*, is lost in evening twilight. *Jupiter* begins retrograde motion in early September. It is in *Aries* at mag. -2.6 to -2.8. *Saturn* (mag. 0.5) is in *Aquarius*, and still retrograding. *Uranus* is in *Aries*, close to Taurus, at mag. 5.9 and *Neptune* is in *Pisces* at mag. 7.8. It comes to opposition on September 19 just south of the asterism known as the 'Circlet' (see the chart on page 177).

Viewing Earth satellites

There are now so many objects orbiting the Earth that if you watch the sky shortly after sunset or before dawn, you are almost certain to see a tiny spot of light crossing the sky. Why 'shortly after sunset or before dawn'? Well, the satellite itself must be illuminated by the Sun, while you are in darkness. Satellites often disappear as they pass into Earth's shadow. Even the massive constellations of satellites that are now being launched and which will seriously jeopardise astronomy in

A sequence of images, taken with a specially equipped telescope (see pages 138 and 139) of the International Space Station as its path took it in front of the Sun.

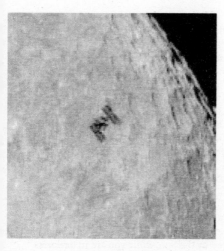

The International Space Station, photographed in front of Mare Serenitatis on the Moon. The large crater to the right of the ISS is Posidonius.

the evening and morning are likely to suffer in the same way. Even those satellites will disappear as they are overtaken by the shadow. Because of their physical composition, sometimes with large flat panels that are highly reflective, satellites may not reflect sunlight evenly in all directions, and many vary in brightness, or even 'flare' as they pass across the sky. The Iridium series of communication satellites are particularly noted for their great variations in brightness.

There are numerous websites that offer details of when satellites may be seen from your location and some are given later under 'Further Information'. With some you may request emails to be sent to you with details of when specific satellites may be visible from your position on Earth.

The largest object in orbit around the Earth is, of course, the International Space Station (ISS). The ISS orbits at an average altitude of 400 kilometres and completes one orbit in about 93 minutes. Once again, you can obtain predictions of where it may be seen from your location. It generally appears as a very bright object, but may occasionally be seen as a dark body passing in front of the Moon.

S

Sunrise and sunset

City	Date	Sunrise	Sunset
Buenos Aires, Argentina			
	Sep. 01	10:13	21:35
	Sep. 30	09:32	21:56
Cape Town, South Africa			
	Sep. 01	05:05	16:28
	Sep. 30	04:25	16:48
London, UK			
	Sep. 01	05:13	18:48
	Sep. 30	05:59	17:42
Los Angeles, USA			
	Sep. 01	13:27	02:20
	Sep. 30	13:47	01:40
Nairobi, Kenya			
	Sep. 01	03:30	15:35
	Sep. 30	03:19	15:26
Sydney, Australia			
	Sep. 01	20:13	07:37
	Sep. 30	19:33	07:57
Tokyo, Japan			
	Sep. 01	20:13	09:09
	Sep. 30	20:35	08:27
Washington, DC, USA			
	Sep. 01	10:37	23:39
	Sep. 30	11:03	22:53
Wellington, New Zealand			
	Sep. 01	18:45	05:56
	Sep. 30	17:56	06:25

NB: the times given are in Universal Time (UT)

The Moon's phases and ages

Northern hemisphere

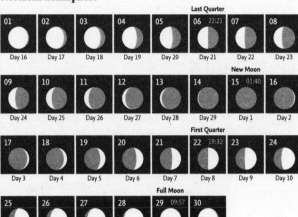

Southern hemisphere

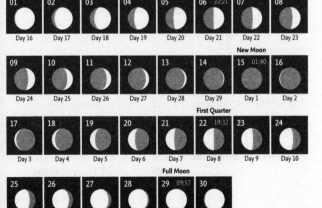

The Moon

On September 4, the waning gibbous Moon is 3.3° north of *Jupiter* (mag. -2.6) in *Aries*. The next day, it is 2.8° north of *Uranus* in *Aries*, which is much fainter at mag. 5.8. By September 6, at Last Quarter, it is 9.3° north of *Aldebaran* in *Taurus*. On September 10, the Moon is 1.6° south of *Pollux*. On September 11, it is 11.4° north of brilliant *Venus* at mag. -4.7 in *Cancer* in the morning sky. On September 13, it is 4.1° north of *Regulus* (mag. 1.4), which may just be visible in morning twilight. By September 16, one day after New Moon, it is 0.7° north of *Mars* in *Virgo*, at the end of evening twilight. The next day it is 2.4° north of *Spica*. By September 21, the Moon is 0.9° north of *Antares* in *Scorpius*. On September 27, the waxing gibbous Moon is 2.7° south of *Saturn* (mag. 0.5) in *Aquarius*.

The Japanese spacecraft *Hayabusa 2* dropped two surface rovers onto the asteroid *Ryugu* on 21 September 2018. A further rover was deployed on 3 October 2018.

Corn Moon

To the peoples of North America, September was particularly important because corn (maize) was ready for harvest. Many tribes had names for the Full Moon that referred to corn, such as 'Corn Maker Moon' among the Abenaki of northern Maine; 'Middle Moon between harvest and eating corn' to the Algonquin in the Northeast and Great Lakes area, although it was the 'Moon when freeze begins on stream's edge' for the Cheyenne of the Great Plains.

In Europe, the September Full Moon was generally called the 'Harvest Moon', and technically this was the first Full Moon after the autumnal equinox (23 September in 2023). In general, this Full Moon comes in September, but approximately once every three years, Full Moon comes in October, when it is that Full Moon that is known as the 'Harvest Moon'. The term 'Harvest Moon' was the only name for the Full Moon that was determined by the equinox, rather than being specific to any particular month. Other names for this Full Moon, from the Old World, and specifically from the Old English/Anglo-Saxon calendar, were 'Full Corn Moon' – in this case 'corn' meaning wheat or barley – as well as 'Barley Moon'.

The surface of Pluto, as imaged by the New Horizons spaceprobe. The large, apparently flat region is known as Sputnik Planitia.

New Horizons

The *New Horizons* spaceprobe was launched in January 2006. The spacecraft was designed to make observations of the planet Pluto during a fly-by on 14 July 2015, and also make observations of more distant objects in the Kuiper Belt beyond the planet. (These objects – the Kuiper Belt Objects (KBO), which include Pluto – are strongly affected by the gravitational influence of Neptune.)

The New Horizons probe was able to obtain images of the surface of Pluto and of its large satellite, Charon. The observations confirmed the existence of a thin atmosphere and atmospheric hazes. Water ice has been detected on parts of the surface and details of the surface have now been confirmed to show convective cells of nitrogen ice.

After its fly-by of Pluto, the spaceprobe was directed to the KBO 486958, now known as Arrokoth.

An image of the Kuiper Belt Object, now known as Arrokoth, as obtained by the New Horizons spaceprobe.

S

Calendar for September

Date	Time	Event
01		α-Aurigid meteor shower maximum
01	07:21	Neptune 1.4°N of the Moon
04	19:47	Jupiter 3.3°S of the Moon
05	08:45	Uranus 2.8°S of the Moon
06	11:09	Mercury at inferior conjunction
06	17:21	Aldebaran 9.3°S of the Moon
06	22:21	Last Quarter
10–Nov.20		Southern Taurid meteor shower
10	04:10	Pollux 1.6°N of the Moon
11	12:59	Venus 11.4°S of the Moon
12	15:43	Moon at apogee = 406,291 km
13	02:39	Regulus 4.1°S of the Moon
13	17:40	Mercury 6.0°S of the Moon
15	01:40	New Moon
16	19:20	Mars 0.7°S of the Moon
17	15:51	Spica 2.4°S of the Moon
19	11:17	Neptune at opposition (mag. 7.8)
21	08:27	Antares 0.9°S of the Moon
22	13:16	Mercury at greatest elongation (17.9°W, mag. -0.5)
22	19:32	First Quarter
23	06:50	September equinox
24		New Zealand Daylight Saving Time begins
27	01:29	Saturn 2.7°N of the Moon
28	00:59	Moon at perigee = 359,911 km
28	16:59	Neptune 1.4°N of the Moon
29	09:57	Full Moon

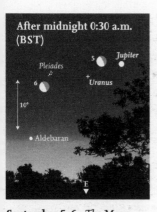

September 5–6 · *The Moon passes Jupiter, Uranus (mag. 5.7) and the Pleiades (as seen from London).*

September 9–11 · *The crescent Moon passes Pollux and Castor and is near Venus on September 11 (as seen from central USA).*

September 21 · *The Moon is close to Antares, between it and Sabik (as seen from Sydney).*

September 26 · *The Moon is below Saturn at a latitude of 35°. Fomalhaut is lower and due south (as seen from central USA).*

S

185

northwest   north   northeast

September – Looking North

For mid-northern observers, **Cassiopeia** is high overhead, and **Cepheus** is 'upside-down', apparently hanging from the Milky Way, high above the Pole. **Perseus** is high in the northeast, at about the same altitude as **Ursa Minor** and **Polaris**.

For those same observers, **Ursa Major** is now 'right way up', low in the south, although some of the southernmost stars are lost along or below the horizon. **Auriga**, with bright **Capella**, and the small triangle of the 'Kids' is clearly visible in the northeast. Higher in the sky, above Perseus, the whole of **Andromeda** is visible, together with the small constellation of **Triangulum** and, still higher, the Great Square of **Pegasus**, and beyond it, most of the zodiacal constellation of **Pisces**.

The clouds of stars forming the Milky Way are not particularly striking in Auriga and Perseus, but beyond Cassiopeia, and on towards **Cygnus**, they become much denser and easier to see.

Cygnus is high in the northwest and **Deneb**, its principal star, is close to the zenith for observers at 40–50° north, with **Lyra** and bright **Vega**, slightly farther west. Farther towards the south, most of **Hercules** is clearly visible, with the 'Keystone' and the globular cluster M13. The head of **Draco** lies between Hercules and Cepheus and the whole of that constellation is easily seen as it curls around Ursa Minor and the North Celestial Pole.

Observers in the far north may be able to detect **Castor** (α Geminorum) skimming the northern horizon, although brighter **Pollux** (β Geminorum) will be too low to be detected until later in the night.

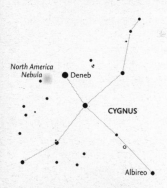

The constellation of Cygnus. Deneb (α Cyg) is one of the stars forming the Summer Triangle. The other two are Vega (α Lyr) and Altair (α Aql).

S

September – Looking South

The three stars forming the apices of the (northern) Summer Triangle, **Deneb** (α Cygni), **Vega** (α Lyrae), and **Altair** (α Aquilae) are prominent in the southwest. The Great Rift is clearly visible, starting near Deneb and running down the centre of the Milky Way towards the centre of the Galaxy in **Sagittarius**, and beyond into **Scorpius**, only petering out in **Centaurus** and **Crux**.

Most of the constellation of **Capricornus** lies just west of the meridian, with **Aquarius** just slightly farther north on the eastern side. The next zodiacal constellation, **Pisces**, is clearly seen, including the prominent asterism, known as the 'Circlet'. Much of the constellation of **Cetus** is visible south of it. South of Aquarius is **Piscis Austrinus**, with its single bright star, **Fomalhaut**, in an otherwise fairly barren area of sky. Still slightly farther south is the undistinguished constellation of **Sculptor**, and, closer to the meridian, the line of stars that forms part of **Grus**.

For observers south of the equator, Sagittarius is high in the west. Below it is the curl of stars that is **Corona Australis**. Below (south of) Sagittarius is the tail and 'sting' of Scorpius. Still farther south is the constellation of **Lupus** and the scattered stars of **Centaurus**, with **Rigil Kentaurus** and **Hadar** (α and β Centauri, respectively). Between the 'sting' of Scorpius and those two bright stars are the small constellations of **Ara** and **Triangulum Australe**. Next comes Crux itself. If the long axis of the cross is extended right across the sky over the largely empty area of sky around the South Celestial Pole, it points in the general direction of **Achernar** (α Eridani). Between Achernar and Ara, west of the meridian, lies the constellation of **Pavo** with it sole bright star **Peacock** (α Pavonis). To the south, the brightest star of the southern hemisphere, **Canopus** (α Carinae) is hugging the horizon for observers at the latitude of Sydney in Australia.

For observers at 30°S, the constellation of Piscis Austrinus, with its single bright star, Fomalhaut, is almost overhead.

On 5 September 1977, the *Voyager 1* spaceprobe was launched. It studied Jupiter, Saturn, and Saturn's satellite, Titan. It became the first object to leave the heliosphere and enter interstellar space.

October

October – Introduction

There is one major meteor shower (the *Orionids*) and three minor showers active during October. The main shower is the Orionids which are fairly reliable in October. Like the May η-*Aquariid* shower, the Orionids are associated with Comet 1P/Halley. During this second pass through the stream of particles from the comet, slightly fewer meteors are seen than in May, but conditions are more favourable for northern observers. In both showers the meteors are very fast, and many leave persistent trains. Although the Orionid maximum is quoted as October 21–22 for 2023, in fact there is a very broad maximum, lasting about a week roughly centred on that date, with hourly rates around 25. Occasionally, rates are higher (50–70 per hour). In 2023, the Moon is a waxing crescent just before First Quarter at the nominal maximum, so conditions are not particularly favourable.

The faint shower of the *Southern Taurids* (often with bright fireballs) peaks on October 10–11. The Southern Taurid maximum occurs between Last Quarter and New Moon, so conditions are favourable. Towards the end of the month (around October 20), another shower (the *Northern Taurids*) begins to show activity, which peaks early in November. The parent comet for both Taurid showers is Comet 2P/Encke, which has the shortest period (3.3 years) of any major comet. The meteors in both Taurid streams are relatively slow and bright.

A short, minor northern shower, the *Draconids*, begins on October 6 and peaks on October 8–9. Although the rate is about 10 meteors per hour, the whole shower occurs at Last Quarter and a few days later in 2023, so observing conditions are reasonable.

The most notable astronomical event of 2023 is the annular eclipse of the Sun on October 14. The path of totality starts at the northwestern coast of the USA, sweeps across to the Gulf of Mexico, then crosses Central America and the north of South America, ending at the coast of Brazil. (See the diagram opposite.)

The planets

Mercury is lost in twilight close to the Sun, and is at superior conjunction on October 20. **Venus**, in **Leo**, is bright (mag. -4.7 to -4.4) in the morning sky and moving closer to the Sun. **Jupiter** is retrograding in **Aries** at mag. -2.8 to -2.9. **Saturn** is also retrograding in **Aquarius**. **Uranus** is still in **Aries** at mag. 5.8 and **Neptune** is slowly retrograding in **Pisces** at mag. 7.8.

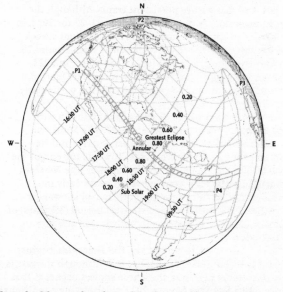

The path of the annular eclipse of October 14 sweeps right across Northern, Central and South America. Maximum eclipse occurs off the coast of Nicaragua.

Sunrise and sunset

City	Date	Sunrise	Sunset
Buenos Aires, Argentina			
	Oct. 01	09:30	21:56
	Oct. 31	08:53	22:22
Cape Town, South Africa			
	Oct. 01	04:24	16:49
	Oct. 31	03:47	17:13
London, UK			
	Oct. 01	06:01	17:39
	Oct. 31	06:52	16:36
Los Angeles, USA			
	Oct. 01	13:48	01:39
	Oct. 31	14:12	01:02
Nairobi, Kenya			
	Oct. 01	03:19	15:26
	Oct. 31	03:12	15:21
Sydney, Australia			
	Oct. 01	19:32	07:57
	Oct. 31	18:55	08:22
Tokyo, Japan			
	Oct. 01	20:36	08:26
	Oct. 31	21:02	07:47
Washington, DC, USA			
	Oct. 01	11:04	22:51
	Oct. 31	11:34	22:09
Wellington, New Zealand			
	Oct. 01	17:55	06:26
	Oct. 31	17:09	07:00

NB: the times given are in Universal Time (UT)

The Moon's phases and ages

Northern hemisphere

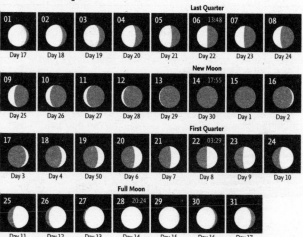

Southern hemisphere

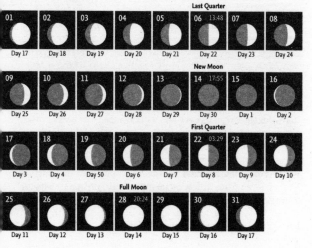

The Moon

On October 2, the Moon passes 3.4° north of *Jupiter* (mag. -2.8), retrograding in *Aries*. Later that day it is 2.9° north of *Uranus* (mag. 5.7). It then passes the *Pleiades* and on October 4 is 9.4° north of *Aldebaran*. By October 7, one day after Last Quarter, it is 1.4° south of *Pollux* in *Gemini*. On October 10, the waning crescent is 4.2° north of *Regulus* in *Leo*. Later the same day it is 6.5° north of brilliant *Venus* (mag. -4.6) also in *Leo*. On October 14, at New Moon, there is an annular solar eclipse with the path running right across North, Central and South America (see pages 20–21, and the map on page 193). Later that day, the Moon is 2.4° north of *Spica* in twilight. By October 18, it is 0.8° north of *Antares* in *Scorpius*. On October 24, the Moon is 2.8° south of *Saturn* in *Aquarius*. On that day there is a minor partial lunar eclipse. By October 29, one day after Full, the Moon is 3.1° north of *Jupiter* and on October 30 it is 2.9° north of *Uranus*, both planets are in *Aries*. The next day, October 31, the waning gibbous Moon passes 9.4° north of *Aldebaran*.

Hunter's Moon

In the northern hemisphere of the Old World, October was the month in which people prepared for the coming winter by hunting wild animals, slaughtering livestock and preserving meat for food. This caused the Full Moon in October to become known as the 'Hunter's Moon'. Every three years, however, the first Full Moon after the autumnal equinox fell, not in September, but early in October, when it was also known as the 'Harvest Moon'. The October Full Moon was also known as the 'Dying Grass Moon' and the 'Blood' or 'Sanguine Moon'. ('Blood Moon' is also the term sometimes applied to the Moon during a lunar eclipse.)

In the New World there was a great variety of names. Many were related to the changes in autumn, such as 'Leaf-falling Moon', 'Falling Leaves Moon' or simply 'Fall Moon'. To the Algonquin of the Northeast and Great Lakes area it was the 'White Frost on Grass Moon'. The Assiniboine of the Northern Plains had a rather different sort of name. To them it was the 'Joins Both Sides Moon'.

Olbers' Paradox

The German astronomer Heinrich Wilhelm Olbers (1758–1840) posed a paradox about the night sky. If the universe were static (unchanging), homogeneous (i.e., the same everywhere) and with a population of an infinite number of stars, then any line of sight must eventually encounter the surface of a star. So the universe should appear uniformly bright wherever one looked and thus be light at night. Obviously one (at least) of the three basic assumptions must be incorrect. The most widely accepted explanation is that the universe is expanding away – or rather, that spacetime is expanding – from its origin in the Big Bang. The expansion causes the redshift of light from any stars or galaxies towards longer wavelengths. Eventually the light becomes invisible to human eyes, hence the darkness at night. This may be taken as evidence that the universe is dynamic, i.e., changing.

The problem of an infinite number of stars in the universe and its significance for the darkness at night was commented upon by a Greek monk, Cosmas Indicopleustes, as long ago as the sixth century, but Olbers was the first to put the problem into a paradoxical form. The paradox is sometimes known as 'the dark night sky paradox'.

There are other theories that account for the darkness of the night sky, without having recourse to a Big Bang from which spacetime is expanding, causing the light from distant stars and galaxies to be shifted into the red; to longer and longer wavelengths, until it becomes invisible to human eyesight and our telescopes. One such theory is the 'Fractal Universe' in which stars occur in a fractal distribution where there are fewer stars, that is, the density of stars decreases as the volume considered increases. This would lead to a dark night sky, but does not rule out that the Universe came into being in a Big Bang.

O

Calendar for October

01		Australian Daylight Saving Time begins
02–Nov.07		Orionid meteor shower
02	03:20	Jupiter 3.4°S of the Moon
02	17:15	Uranus 2.9°S of the Moon
04	01:46	Aldebaran 9.4°S of the Moon
06–10		Draconid meteor shower
06	13:48	Last Quarter
07	11:02	Pollux 1.4°N of the Moon
08–09		Draconid meteor shower maximum
10		Southern Taurid meteor shower maximum
10	03:42	Moon at apogee = 405,426 km
10	09:20	Regulus 4.2°S of the Moon
10	09:44	Venus 6.5°S of the Moon
14	09:33	Mercury 0.7°N of the Moon
14	17:55	New Moon
14	17:59	Annular solar eclipse
14	22:07	Spica 2.4°S of the Moon
15	16:17	Mars 1.0°N of the Moon
18	13:53	Antares 0.8°S of the Moon
20–Dec.10		Northern Taurid meteor shower
20	05:38	Mercury at superior conjunction
21		Orionid meteor shower maximum
22	03:29	First Quarter
23	23:14	Venus at greatest elongation (46.4°W, mag. -4.5)
24	07:56	Saturn 2.8°N of the Moon
26	01:23	Neptune 1.5°N of the Moon
26	03:02	Moon at perigee = 364,872 km
28	20:13	Partial lunar eclipse
28	20:24	Full Moon
29	01:00	European Daylight Saving Time ends
29	08:14	Jupiter 3.1°S of the Moon
29	17:00 *	Mars (mag. 1.5) 0.4°N of Mercury (mag. -1.0)
30	01:53	Uranus 2.9°S of the Moon
31	11:28	Aldebaran 9.4°S of the Moon

These objects are close together for an extended period around this time.

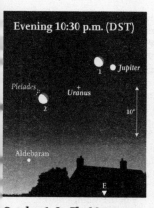

October 1–2 • *The Moon passes Jupiter, Uranus (mag. 5.7) and the Pleiades (as seen from central USA).*

October 10 • *The Moon lines up with Regulus and Venus, with Algieba to its left (as seen from London).*

October 23–24 • *The Moon passes Saturn. Fomalhaut is closer to the horizon and almost due south (as seen from London).*

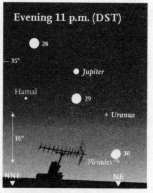

October 28–30 • *The Moon passes Jupiter, Uranus (mag. 5.6) and the Pleiades (as seen from Sydney).*

O

October – Looking North

The Milky Way arches across the northern sky, running from **Auriga** in the east to **Cygnus** and **Aquila** in the west. The constellations of **Perseus**, **Cassiopeia**, and **Cepheus** lie along it. For observers in middle northern latitudes, these constellations are high overhead. (To observers in the far north, Cassiopeia is near the zenith.) **Andromeda** is high in the north, beyond Perseus and the other side of the Milky Way. The two small constellations of **Triangulum** and **Aries** are to the south of it. In the northwest, the constellation of **Lyra** lies farther south, clear of the Milky Way.

Between Cassiopeia and Cygnus lies the zig-zag of faint stars that form the small constellation of **Lacerta**, which is often difficult to recognize because it lies across the Milky Way. Two of the stars forming the Summer Triangle, **Deneb** (α Cygni) and **Vega** (α Lyrae) are readily visible, but the third, **Altair** (α Aquilae), is approaching the northwestern horizon. So too, much farther north, is the constellation of **Hercules**. The head of **Draco** is at about the same altitude as **Polaris** in **Ursa Minor**, as is **Capella** (α Aurigae). The large constellation of **Ursa Major** is now directly beneath the Pole, visible above the horizon to the north. **Castor** and **Pollux** (α and β Geminorum, respectively) are in the northeast, while the northernmost stars of **Boötes** and the circlet of **Corona Borealis** are in the northwest. The brightest star in Boötes, orange-tinted Arcturus, is below the horizon early in the night.

The long chain of faint stars that is the constellation of **Lynx** runs almost vertically between Ursa Major and **Gemini**, with the other faint constellation of **Camelopardalis** between it and Perseus.

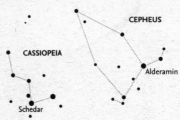

This month two of the northern circumpolar constellations, Cassiopeia and Cepheus, are almost at the meridian.

O

southeast   south   southwest

October – Looking South

The Great Square of **Pegasus** is now on the meridian due north, with the constellation of **Andromeda** stretching away from it in the northeast. The star at the northeastern corner of the Great Square is actually **Alpheratz** (α Andromedae). The two parts of the zodiacal constellation of **Pisces** are to the south and east of Pegasus. **Aquarius** and the non-zodiacal constellation of **Cetus** are slightly farther south while **Capricornus** is sinking in the west.

South of Aquarius is **Pisces Austrinus** with its single bright star, **Fomalhaut**, while straddling the meridian slightly east of it is the faint constellation of **Sculptor**. Below (south of) these two constellations are the roughly cross-shaped constellation of **Grus** with, on the other side of the meridian, **Phoenix**. Just below the constellation of Phoenix is **Achernar** (α Eridani), the bright star that ends the long winding constellation of **Eridanus** that actually begins near Rigel in Orion. Below Achernar is the triangular constellation of **Hydrus**, with the **Small Magellanic Cloud** (SMC) on its northwestern side. At about the same altitude towards the west is the constellation of **Pavo** with **Peacock** (α Pavonis).

The **Large Magellanic Cloud** (LMC) lies southeast of Hydrus, partly in each of the two faint constellations of **Dorado** and **Mensa**. The Milky Way lies in the east, where portions of **Sagittarius** and **Scorpius** are clearly visible. Roughly level with the 'Sting' of Scorpius are the constellations of **Ara** and **Triangulum Australe**. Farther east are the undistinguished constellations of **Apus**, **Chamaeleon** and **Volans**. Even farther to the east is brilliant **Canopus** (α Carinae), the second brightest star in the sky (after Sirius) at mag. -0.6.

For observers at about 30° south (roughly the latitude of Sydney in Australia), the two brightest stars of **Centaurus** (**Rigil Kentaurus** and **Hadar**) are brushing the horizon, while the small constellation of **Crux** is lost below it. Only observers even farther south will be able to see that constellation in full, together with the constellation of **Lupus**, the full extent of Centaurus and the large constellation of **Vela**, with **Puppis** in the southeast.

The Voyager probes

Following the success of the Pioneer spaceprobes, NASA designed the Voyager program to take advantage of the favourable positions of the planets Jupiter and Saturn. The two spacecraft were launched in 1977, **Voyager 2** being actually the first to be launched. **Voyager 1** was set on a faster track, primarily so that it could make observations of Saturn's satellite Titan, which would be favourably placed. It made gravity-assist passes of Jupiter and Saturn. (These manoeuvres actually took it out of the plane of the ecliptic, so it could not continue on to make observations of any other objects.) Both spaceprobes made successful observations of Jupiter and Saturn, with the data being sent back to Earth. The trajectory for **Voyager 2** (launched first) was designed so that the probe could be sent on to Uranus and Neptune. It also obtained gravity assists from the two gas giants to enable it to travel onwards to the outer planets.

Miranda, the satellite of Uranus as viewed by Voyager 2. The 'racetrack-like' feature on the left is known as a corona. Also visible is a feature now known as 'The Chevron' and, on the limb, the cliffs of Verona Rupes.

The immensely high icy scarp of Verona Rupes. These cliffs appear to be 20 km in height, making them the highest in the Solar System.

The spacecraft obtained many observations of Jupiter's highly complex meteorology. They observed Jupiter's main satellites (the Galilean moons) in detail and, in doing so, discovered the volcanic activity on Io, which has proved to be the most volcanically active object in the Solar System. At Saturn, apart from Voyager's first observations of Titan, they carried out extensive observations of the ring system, discovering the existence of numerous 'ringlets', 'braided' rings, and how the numerous satellites influence the nature of the rings. This included the discovery of 'shepherd' satellites that govern the structure of some of the individual rings.

Voyager 2 is the only spacecraft to have visited the two outermost gas giant (more properly, the 'ice giant') planets. It flew by Uranus in 1986 and Neptune in 1989. At Uranus it found a very extensive magnetic field, and no less than 10 additional satellites. It provided dramatic images of Miranda, the innermost of the large satellites. Miranda has an extremely varied topography with some unique features. It also exhibits what is probably the highest cliff in the Solar System, known as Verona Rupes, which is believed to be 20 km high. Miranda, like the other satellites of Uranus, orbits the planet in its equatorial plane, which is highly inclined to the ecliptic. (The axial tilt is 97.77°.)

O

At Neptune, *Voyager 2* observed a 'Great Dark Spot' in the southern hemisphere, somewhat comparable to Jupiter's Great Red Spot. (Similar features, including major storms have since been observed by the Hubble Space Telescope in the northern hemisphere.) The spacecraft also revealed that Neptune has a system of narrow rings – thought, before the encounter, to be 'arcs' rather than true rings. (The nine rings forming the ring system around Uranus had been known since an occultation on 10 March 1977.) The planet proved to have a more complex and dynamic meteorology than Uranus, with fast-moving clouds. Some of the wind systems reach the highest speeds observed in the Solar System, achieving velocities of almost 2,200 km/h.

The two *Voyager* spacecraft have subsequently continued their journeys outwards, and have both now passed the heliopause, the theoretical boundary where the solar wind is halted by the interstellar medium. This is at a distance of about 120 AU.

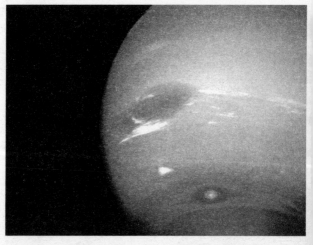

Neptune as viewed by Voyager 2, showing the Great Dark Spot, bordered by high clouds; the fast-moving white patch of cloud, known as 'Scooter'; and the Small Dark Spot.

November

November – Introduction

Three meteor streams begin in earlier months, but continue into November. The **Southern Taurid** shower, which began as long ago as September 10, may continue until November 20. Similarly, the **Orionids**, which began on October 2, continue until November 7. (Both of these showers are described in earlier months.) The third of these enduring meteor streams, the **Northern Taurid** shower, begins in late-October (October 20), and reaches its maximum – although with just a low rate of about five meteors per hour – on November 12. This is one day before New Moon, so conditions are particularly favourable to see these meteors. The shower gradually tails off, ending around December 10. There is an apparent 7-year periodicity in fireball activity, but 2023 is unlikely to be a peak year.

Far more striking, however, is the major shower of the **Leonids**, which have a relatively short period of activity (November 6–30), with maximum on November 17–18. This shower is associated with Comet 55P/Tempel-Tuttle, which has an orbital period of 33.22 years. The shower has shown extraordinary activity on various occasions with many thousands of meteors per hour. High rates were seen in 1999, 2001 and 2002 (reaching about 3000 meteors per hour) but have fallen dramatically since then. The rate in 2023 is likely to be about 15 per hour. Because of the radiant's location this shower is best seen from the northern hemisphere. These meteors are the fastest shower meteors ever recorded (about 70 km per second) and often leave persistent trains. Apart from the sheer numbers occasionally seen, the shower is very rich in faint meteors. In 2023, maximum is when the Moon is a waxing crescent (days 4–5 of the lunation), so conditions are reasonably favourable.

There is an enigmatic, minor southern meteor shower that begins activity in late November (nominally November 28). This is the **Phoenicids**. Little is known of this shower, partly because the parent comet is believed to be the disintegrated comet D/1819 W1 (Blanpain). With no accurate knowledge of the location of the remnants of the comet, predicting the possible rate becomes little more than guesswork. The rate is variable and may rapidly increase if the orbit is nearby. There is a tendency for bright meteors to be frequent and all the meteors are fairly slow. The radiant is located within the constellation of Phoenix, not far from the border with Eridanus and the bright star *Achernar* (α Eridani).

The planets

Mercury is in evening twilight, but may become visible at the end of the month on the border of *Ophiuchus* and *Sagittarius* at mag. -0.4 to -0.5. *Venus* (mag. -4.4 to -4.2) begins in *Leo* and moves across *Virgo*. *Mars* is lost in twilight near the Sun. *Jupiter* (mag. -2.9) comes to opposition in *Aries* on November 3 (see the map op page 213). *Saturn* (mag. 0.7 to 0.8) initially retrograding in *Aquarius*, resumes direct motion on November 5. *Uranus* (mag. 5.6) is at opposition in *Aries* on November 13 (see the map below) and *Neptune* is slowly retrograding in *Pisces* at mag. 7.9.

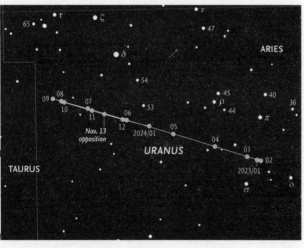

The path of Uranus in 2023. Uranus comes to opposition on November 13 in Aries. Stars down to magnitude 7.5 are shown.

N

Sunrise and sunset

City	Date	Sunrise	Sunset
Buenos Aires, Argentina			
	Nov. 01	08:52	22:23
	Nov. 30	08:34	22:51
Cape Town, South Africa			
	Nov. 01	03:46	17:14
	Nov. 30	03:28	17:41
London, UK			
	Nov. 01	06:54	16:34
	Nov. 30	07:43	15:56
Los Angeles, USA			
	Nov. 01	14:13	01:01
	Nov. 30	14:40	00:44
Nairobi, Kenya			
	Nov. 01	03:12	15:21
	Nov. 30	03:16	15:27
Sydney, Australia			
	Nov. 01	18:54	08:23
	Nov. 30	18:37	08:50
Tokyo, Japan			
	Nov. 01	21:03	07:46
	Nov. 30	21:31	07:28
Washington, DC, USA			
	Nov. 01	11:35	22:08
	Nov. 30	12:07	21:47
Wellington, New Zealand			
	Nov. 01	17:07	07:01
	Nov. 30	16:43	07:36

NB: the times given are in Universal Time (UT)

The Moon's phases and ages

Northern hemisphere

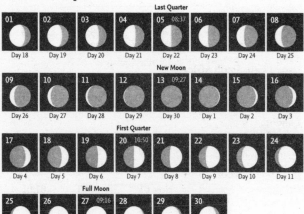

Southern hemisphere

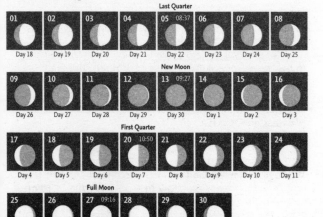

The Moon

On November 3, the waning gibbous Moon is 1.5° south of *Pollux* in *Gemini*. On November 6 it passes 4.2° north of *Regulus*, between it and *Algieba* (γ Leonis). By November 9 it is 1.0° north of *Venus* (mag. -4.4), which is then in *Virgo*. On November 11 the Moon passes 2.4° north of *Spica* in Virgo. By November 20, the Moon is 2.7° south of *Saturn* in *Aquarius*. On November 25 it passes 2.8° north of *Jupiter* in *Aries*, then the next day, 2.6° north of *Uranus* (mag. 5.6) also in Aries. On November 27 it is 2.6° north of *Aldebaran* in *Taurus*.

Beaver Moon

In North America, the Full Moon of November has come to be called 'Beaver Moon', because beavers become particularly active at this time, preparing their lodges and food supplies for winter. But this applies to a few only of the North-American tribes. Most see it as marking the beginning of heavy frosts, with names such as 'Freezing River Maker Moon', 'Freezing Moon', 'Rivers Begin to Freeze Moon', and 'Frost Moon', although to the Haida of Alaska, where snow lies for many months, it was the 'Snow Moon'.

The Japanese spacecraft *Hayabusa 2* left the asteroid *Ryugu* on 13 November 2019, returning its samples to Earth in December 2020.

In the Old World it was sometimes known as the 'Frosty Moon' or even, occasionally, as the 'Oak Moon', although this last term was more often applied to the Full Moon in December. If it was the last Full Moon before the winter solstice it was also sometimes known as the 'Mourning Moon'.

The Kuiper Belt

Beyond the orbit of Neptune, there is a region occupied by a large number of small bodies. (These bodies are collectively known as 'Trans-Neptunian Objects' or TNOs.) This belt is similar to the main asteroid belt between Mars and Jupiter, but the Kuiper Belt is much wider, extending from about 30 AU to 50 AU. The objects within it (Kuiper Belt Objects or KBOs) are also generally more massive than those in the inner belt. The orbits are dynamically stable. Both belts are thought to contain small bodies that are remnants left over when the Solar System was formed. Those in the inner asteroid belt are generally of rocky material, whereas those in the Kuiper Belt are icy. In this context, 'ices' includes water ice and frozen methane and ammonia. It is believed that most of the dwarf planets are found in this region, Pluto being the largest such body. Other known dwarf planets are Eris, Haumea, Makemake, Orcus, and Quaoar.

It was originally thought that periodic comets originated within the Kuiper Belt. However, it has now been established that comets originate in what is known as the 'scattered disc', originally created by the outward motion of Neptune during the early stages of the formation of the Solar System. The orbit of bodies in the scattered disc are dynamically unstable, in contrast to those in the Kuiper Belt itself. Objects in the scattered disc may orbit as far out as 100 AU. Eris, the largest known dwarf planet, has an extremely eccentric orbit, between about 38 and 97 AU. Long-period comets are believed to originate even farther out, in the Oort Cloud (page 227), which surrounds the Sun at much greater distances of 200 to 200,000 AU.

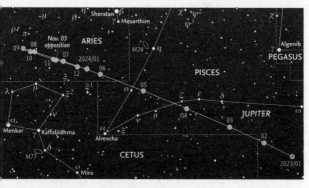

The path of Jupiter in 2023. Jupiter comes to opposition on November 3 in Aries. Stars down to magnitude 6.5 are shown.

N

Calendar for November

03	05:02	Jupiter at opposition (mag. -2.9)
03	19:09	Pollux 1.5°N of the Moon
05	08:37	Last Quarter
06–30		Leonid meteor shower
06	16:59	Regulus 4.2°S of the Moon
06	21:49	Moon at apogee = 404,569 km
09	09:30	Venus 1.0°S of the Moon
11	05:48	Spica 2.4°S of the Moon
12–13		Northern Taurid meteor shower maximum
13	09:27	New Moon
13	13:32	Mars 2.5°N of the Moon
13	17:21	Uranus at opposition (mag. 5.6)
14	14:39	Mercury 1.7°N of the Moon
14	20:18	Antares 0.9°S of the Moon
17–18		Leonid meteor shower maximum
18	05:41	Mars at superior conjunction
20	10:50	First Quarter
20	14:06	Saturn 2.7°N of the Moon
21	21:01	Moon at perigee = 369,818 km
22	07:45	Neptune 1.5°N of the Moon
25	11:14	Jupiter 2.8°S of the Moon
26	09:19	Uranus 2.6°S of the Moon
27	09:16	Full Moon
27	21:03	Aldebaran 9.3°S of the Moon
28–Dec.09		Phoenicid meteor shower

November 4 • *Shortly before Last Quarter, the Moon is close to Pollux and Castor (as seen from Sydney).*

November 7 • *The Moon is between Regulus and Algieba (as seen from Sydney).*

November 9–11 • *The Moon is close to Venus on November 9. Two days later it passes Spica (as seen from London).*

November 25–27 • *The Moon passes Jupiter, Uranus (mag. 5.6) and the Pleiades (as seen from central USA).*

N

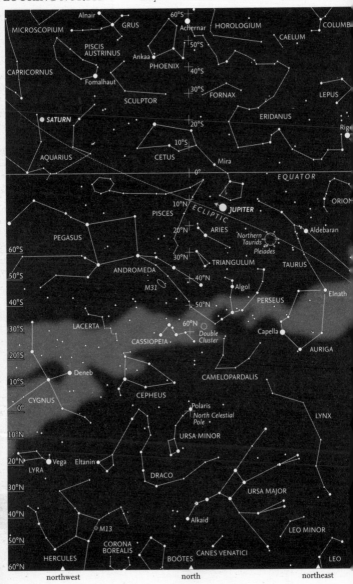

November – Looking North

For observers at mid-northern latitudes, the Milky Way is now high in the north and northwest. **Cassiopeia** is not far from the zenith and **Cepheus** is 'on its side' in the northwest. **Capella** (α Aurigae) and the whole constellation of **Auriga** are high in the northeast. One of the stars of the Summer Triangle, Altair in Aquila, is now disappearing below the horizon. The other two: **Deneb** in **Cygnus** and **Vega** in **Lyra** are still visible, but low towards the northwestern horizon. **Eltanin** (γ Draconis) the brightest star in the 'head' of **Draco**, is at about the same altitude as Vega. **Ursa Minor**, together with **Polaris** itself, is slightly higher in the sky.

The end of the 'tail' of **Ursa Major** (that is, the star **Alkaid**, η Ursae Majoris), is almost due north, although the body of the constellation has begun to swing round into the northeast. Most of the fainter stars in the constellation to the south and east are easily visible, as is the undistinguished constellation of **Canes Venatici** to the southwest of it. Another insignificant constellation, often ignored, is **Leo Minor** to the southeast of Ursa Major. This consists of little more than three faint stars. **Gemini**, with the two bright stars, **Castor** and **Pollux**, is off to the east, with the line of stars forming **Lynx** between that constellation and Ursa Major. To the west of Ursa Major is **Hercules** although some of its stars are very low, close to the horizon, and likely to be difficult to see. Slightly towards the south are the northernmost stars of **Boötes**, although Arcturus itself is below the northern horizon.

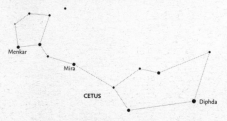

This time of the year, the constellation of Cetus and its famous variable star, Mira (o Ceti), is clearly visible. Because the celestial equator crosses the constellation, it is visible from almost every latitude.

N

southeast   south   southwest

November – Looking South

Several major constellations dominate the southern sky. There is *Pegasus* with the Great Square and *Andromeda* above and to the east of it. *Pisces* straddles the meridian, with the 'Circlet' to the south of the Great Square. The zodiacal constellation of *Aquarius* is sinking in the west, but is still clearly visible. To the east is *Taurus* and orange-tinted *Aldebaran*. Below (south) of Pisces the whole of *Cetus* and its famous variable star, *Mira* (o Ceti), is clearly visible. *Orion* is rising in the east, and the beginning of *Eridanus* at the stars λ Eridani and *Cursa* (β Eridani) near *Rigel* (β Orionis) is visible as are most of the stars in the long chain that forms this constellation as it winds its way south. The brightest star, *Achernar* (α Eridani), at the very end of the constellation, is on the horizon for observers at a latitude of 30° north. This constellation once ended at *Acamar*, the moderately bright star south of the small constellation of *Fornax*, and originally called Achernar until the constellation was extended and the name transferred to the current, brighter star.

Immediately north and west of Achernar is *Phoenix* and between that and Aquarius lie the two constellations of *Sculptor* and *Piscis Austrinus*, the latter with its solitary bright star, *Fomalhaut*. Below Piscis Austrinus is the constellation of *Grus*. Both the Magellanic Clouds are clearly seen. The Small Cloud (*SMC*) on the border between *Tucana* and *Hydrus* is almost on the meridian, and the Large Cloud (*LMC*) is farther east, bordered by *Dorado* and *Mensa*. *Canopus* (α Carinae) is at almost the same altitude, and the stars of *Puppis* are farther east. South of Canopus are the stars of *Vela* and *Carina*. *Crux* is now more-or-less 'upright', just east of the meridian. *Rigil Kentaurus* and *Hadar* are on the other side of the meridian. All these significant stars are right on the horizon for observers at 30° south, and only visible later in the night and later in the month. The zodiacal constellation of *Scorpius* is disappearing in the southwest and the stars of the sprawling constellation of *Centaurus* and those of *Lupus* are low in the sky.

N

Pioneer 10

On 2 March 1972, NASA launched the *Pioneer 10* spaceprobe. It was the very first spacecraft destined to study the objects in the outer Solar System, and became the first object to fly past Jupiter, at a closest distance of 132,252 km on 6 November 1973. One of its images is shown here.

An image of Jupiter obtained by Pioneer 10 in November 1976. The Great Red Spot is just visible on the eastern limb, with a white oval farther west.

Pioneer 10 passed the orbit of Saturn in 1976, that of Uranus in 1979 and Neptune's orbit on 13 June 1983.

Pioneer 10 became the first spaceprobe to leave the Solar System. Its weak signal was used for training purposes after March 1997 and the mission officially ended that month. The last useful telemetry from the instruments on board came in April 2002. Communications with the probe were lost on 23 January 2003, when the signal was extremely weak and unable to provide any usable data. The signal was finally lost because of a loss of power to the radio transmitter.

December

December – Introduction

The last major meteor shower of the year is the *Geminid* shower, which is visible over the period December 4–20 and comes to maximum on December 14–15. In 2023 the Moon is just after New Moon, so conditions are very favourable. It is one of the most active showers of the year, and in some years is the strongest, with a peak rate of around 100 meteors per hour. It is the one major shower that shows good activity before midnight. The meteors have a much higher density than most meteors (which are derived from cometary material). It was eventually established that the *Geminids* and the asteroid Phaethon had similar orbits. (Phaethon is the named asteroid that comes closest to the Sun at perihelion. There are unnamed numbered objects with closer perihelia.)

The Geminids are assumed to consist of denser, rocky material shed from Phaethon. They are slower than most other meteors and often seem to last longer. The brightest frequently break up into numerous luminous fragments that follow similar paths across the sky.

There is a second, minor, northern shower: the *Ursids*, which is active December 17–26, peaking on December 22–23. The maximum rate is generally approximately 10 meteors per hour, although much higher rates have been observed. The radiant is in Ursa Minor, near the star Kochab (β UMi). The stream is associated with Comet 8P/Tuttle, and is an extremely dense cluster of particles.

There are two southern meteor showers that may be seen in December. The *Phoenicid* shower begins in late November (November 28), and continues into December, reaching its weak maximum on December 2. The other shower, the *Puppid Velids*, has its radiant on the border between the two constellations of Puppis and Vela. The shower begins on December 1, lasting until December 15, with maximum on December 7. (In 2023, this is two days after Last Quarter.) It is a weak shower with a maximum rate of about 10 meteors per hour, although bright meteors are frequently visible.

The planets

Mercury reaches greatest elongation east in the evening sky (21.3°, mag. -0.5) on December 4. *Venus* (mag. -4.1) becomes visible in the morning sky, later in the month. *Mars* is initially in *Libra* and moves across *Scorpius* into *Ophiuchus* during the month. *Jupiter* (mag. -2.8 to -2.6) is moving westwards (with retrograde motion) in *Pisces. Saturn* is in *Aquarius* at mag. 0.9. *Uranus* is still retrograding in *Aries*, and is mag. 5.7 to the end of the year. *Neptune* remains in *Pisces* on the border with *Aquarius* at mag. 7.9, resuming direct motion on December 12. Minor planet *(4) Vesta* comes to opposition on December 21, at mag. 6.4, in Orion (see the maps below).

A finder chart for the position of Vesta at its opposition on December 21. The grey area is shown in more detail on the chart below, showing stars down to magnitude 7.0.

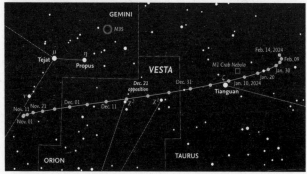

Sunrise and sunset

City	Date	Sunrise	Sunset
Buenos Aires, Argentina			
	Dec. 01	08:34	22:51
	Dec. 31	08:43	23:09
Cape Town, South Africa			
	Dec. 01	03:28	17:42
	Dec. 31	03:37	18:00
London, UK			
	Dec. 01	07:44	15:55
	Dec. 31	08:07	16:01
Los Angeles, USA			
	Dec. 01	14:41	00:44
	Dec. 31	14:59	00:53
Nairobi, Kenya			
	Dec. 01	03:16	15:27
	Dec. 31	03:30	15:41
Sydney, Australia			
	Dec. 01	18:37	08:51
	Dec. 31	18:47	09:09
Tokyo, Japan			
	Dec. 01	21:32	07:28
	Dec. 31	21:50	07:37
Washington, DC, USA			
	Dec. 01	12:08	21:47
	Dec. 31	12:27	21:56
Wellington, New Zealand			
	Dec. 01	16:42	07:37
	Dec. 31	16:51	07:57

NB: the times given are in Universal Time (UT)

The Moon's phases and ages

Northern hemisphere

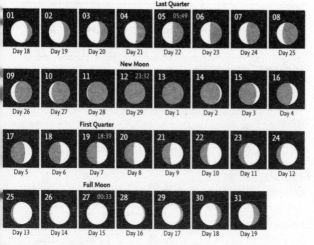

Southern hemisphere

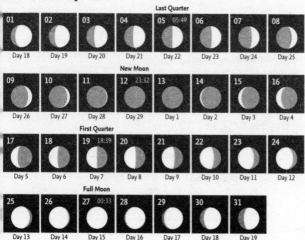

The Moon

On December 1, the Moon is 1.6° south of **Pollux** in **Gemini**. On December 4 the Moon passes 4.0° north of **Regulus**, between it and **Algieba**. By December 8 the waning crescent Moon is 2.3° north of **Spica**. The next day it is 3.6° south of **Venus**. By New Moon on December 12, lost in twilight, it is 0.9° north of **Antares** and later that day, 3.6° south of **Mars**. On December 17 the Moon is 2.5° south of **Saturn** in **Aquarius**. By December 22 the waxing gibbous Moon is 2.6° north of **Jupiter** in **Aries**. It is 2.8° north of **Uranus** (mag. 5.7), also in Aries, the next day. It passes 9.4° north of **Aldebaran** in **Taurus** on December 25. One day after Full Moon, on December 28, it is 1.7° south of **Pollux** and on December 31, 3.8° north of **Regulus**, once again between it and Algieba.

Cold Moon

In the northern hemisphere, the cold of winter begins to dominate life during December. Many of the names for the Full Moon in December on both sides of the Atlantic refer to the temperature, with terms such as 'Cold Moon', 'Winter Maker Moon' and 'Snow Moon' used in North America. The Cree of Canada had a rather strange name: 'Moon when the Young Fellow Spreads the Brush'. Among the Zuñi of New Mexico it was the 'Moon when the Sun has travelled home to rest'. In Europe it was sometimes called the 'Moon before Yule' or the 'Wolf Moon', although that term was more commonly applied to the Full Moon in January.

The Japanese spacecraft **Hayabusa 2** returned samples of the asteroid **Ryugu** to Earth on 6 December 2020. The probe has since been redirected to asteroid 1998 KY_{26}.

Next page: *Comet C/2006 P1 McNaught, imaged on 20 January 2007, from Lawlers Gold Mine, Western Australia (Photographer: Sjbmgrtl).*

The Oort Cloud
The outermost region of the Solar System is believed to consist of a 'cloud' of icy bodies (planetesimals) that were formed early in the history of the Solar System. It is thought that these planetesimals, although formed closer to the Sun, were scattered outwards by the various gravitational effects of the early giant planets. The cloud consists of two distinct regions: an inner disc ('the inner Oort Cloud'), also sometimes known as the 'Hills Cloud', and a spherical 'outer Oort Cloud'. Both regions lie well beyond the heliosphere (the region dominated by the solar wind) and are actually considered to lie in interstellar space. (The two *Voyager* probes [pages 204–206] are now just outside the heliosphere.) The limits of the Oort Cloud are thought to be about 2000 to 200,000 AU. The two regions populated by TNOs (Trans-Neptunian Objects): the Kuiper Belt and the scattered disc (see pages 212 and 229) are much closer to the Sun at about one-thousandth of the distance.

D

Although most of the short-period comets are thought to come from the scattered disc, a few may originate in the Oort Cloud itself. Some comets from the scattered disc may evolve into what are termed 'Centaurs', which may themselves then evolve into short-period comets. Other comets, especially those with extremely long periods (thousands or millions of years) are believed to originate more or less directly from the outer Oort Cloud. It is suggested that tidal effects exerted by the Milky Way galaxy itself are more than sufficient to alter the orbits of objects in the Oort Cloud and cause them to 'fall' in towards the Sun as long-period comets. It is also possible that 'passing' stars may have the same effect.

Dwarf planets

One of the main reasons for the 'demotion' in 2006 (as some might call it) of Pluto from being called a planet to a dwarf planet was the discovery of many, rather similar, large bodies orbiting in the outer Solar System. These Trans-Neptunian Objects (TNOs) include the bodies now known as Eris, Haumea, Makemake, Orcus, Quaoar, Sedna and others. Pluto is the largest of these objects. The most massive is Eris, which is about 27 per cent more massive than Pluto, although not so great in diameter. (Eris is believed to be 2326 km in diameter, as against 2376 km for Pluto.) Eris has one known satellite, Dysnomia, compared with five for Pluto.

The Chinese lunar probe *Chang-e 5* landed in Oceanus Procellarum on 1 December 2020. It returned lunar samples to Earth on 16 December 2020, showing that lunar volcanism had persisted for much later than previously believed.

Plutinos

Plutinos are a group of Trans-Neptunian objects (TNOs) that have a 2:3 orbital resonance with Neptune. Such a resonance (where Neptune completes two orbits, whilst the plutino completes exactly three) is extremely stable. Pluto itself is the largest body in this group. They form part of the Kuiper Belt Objects (KBOs) that orbit outside the orbit of Neptune. All have reasonably stable orbits (in contrast to objects in the 'scattered disc'), but those of the plutinos are stabilized more strongly by the orbital resonance. It has, however, been recently suggested that some orbits may be perturbed by Pluto itself. Apart from Pluto, two of the major plutinos are Orcus and Ixion.

The scattered disc

There are numerous objects in the outer Solar System that do not lie in the same plane as all the major planets. These have highly inclined orbits and are generally known as 'scattered-disc objects' (SBOs). (They are unlike the group of objects known as 'plutinos', of which Pluto is the largest and most significant. Plutinos have stable orbits, with relatively low inclinations and a 2:3 resonance with the orbit of Neptune.) Objects in what is known as the 'scattered disc' not only have inclined orbits, but they also have extremely high eccentricities. Eris, the most massive of these bodies with a mass about 27 per cent greater than Pluto, has a perihelion distance of about 37.9 AU and aphelion of 97.5 AU. It has an orbital period of some 559 years. Its inclination to the ecliptic is 44 degrees.

It is believed that all the bodies in the scattered disc were 'scattered' by gravitational forces during the early stages of the formation of the Solar System, in particular by gravitational effects caused by the outward migration of Neptune. Some have acquired stable orbits (the Kuiper Belt Objects, KBOs), whereas others were sent into highly inclined, eccentric orbits and became Scattered Disc Objects (SBOs).

D

Calendar for December

01–15		Puppid Velid meteor shower
01	04:01	Pollux 1.6°N of the Moon
02		Phoenicid meteor shower maximum
04–20		Geminid meteor shower
04	01:18	Regulus 4.0°S of the Moon
04	14:28	Mercury at greatest elongation (17.9°E, mag. -0.5)
04	18:42	Moon at apogee = 404,346 km
05	05:49	Last Quarter
07		Puppid Velid meteor shower maximum
08	14:45	Spica 2.3°S of the Moon
09	16:53	Venus 3.6°N of the Moon
12	04:52	Antares 0.9°S of the Moon
12	10:55	Mars 3.6°N of the Moon
12	23:32	New Moon
14–15		Geminid meteor shower maximum
14	05:19	Mercury 4.4°N of the Moon
16	18:53	Moon at perigee = 367,901 km
17–26		Ursid meteor shower
17	22:01	Saturn 2.5°N of the Moon
19	13:16	Neptune 1.3°N of the Moon
19	18:39	First Quarter
21	18:56	Minor planet (4) Vesta at opposition (mag. 6.4)
22–23		Ursid meteor shower maximum
22	03:27	December solstice
22	14:24	Jupiter 2.6°S of the Moon
22	18:54	Mercury at inferior conjunction
22	23:26	Minor planet (9) Metis at opposition (mag. 8.4)
23	14:54	Uranus 2.8°S of the Moon
25	05:01	Aldebaran 9.4°S of the Moon
27	00:33	Full Moon
28	03:00 *	Mercury (mag. 2.0) 3.6°N of Mars (mag. 1.4)
28	12:30	Pollux 1.7°N of the Moon
31	09:33	Regulus 3.8°S of the Moon

* These objects are close together for an extended period around this time.

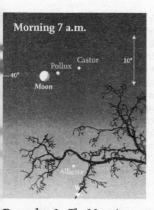

December 1 · *The Moon is lining up with Pollux and Castor, shortly before sunrise (as seen from London).*

December 8–9 · *The Moon passes Spica and the brilliant Venus, mag. -4.0 (as seen from central USA).*

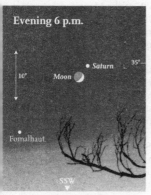

December 17 · *The Moon is between Saturn and the slightly fainter Fomalhaut (as seen from central USA).*

December 22–24 · *In the north, the Moon passes Jupiter, Uranus (mag. 5.7) and the Pleiades (as seen from Sydney).*

D

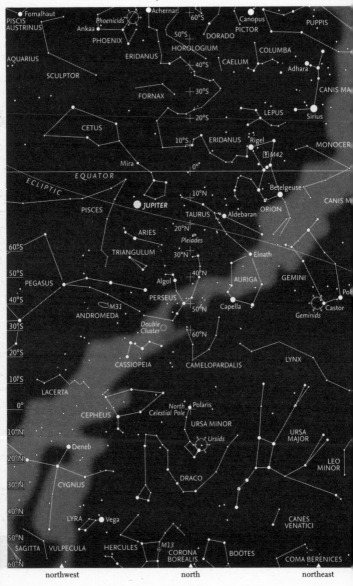

northwest   north   northeast

December – Looking North

For observers close to the equator, *Orion* is high overhead, to the east of the meridian, with *Taurus*, *Auriga* and *Gemini* way above the northern horizon. For observers at mid-northern latitudes, it is *Perseus* that is at the zenith, with *Andromeda* stretching off to the west and the Great Square of *Pegasus* in the northwest. Auriga with *Capella* is slightly to the east. Even farther east are the two bright stars of *Gemini*, *Castor* and *Pollux*.

The Milky Way runs from high in the northeast down to the northwest. Below Perseus is the distinctive shape of *Cassiopeia*, which lies within the Milky Way, and farther down is the zig-zag constellation of *Lacerta*, which is like *Cepheus*, in that both of them are partly within the band of stars. Even farther towards the northwest is *Cygnus* and the beginning of the Great Dark Rift near *Deneb*. The constellation of *Lyra*, with brilliant *Vega*, lies towards the meridian, away from the star clouds of the Milky Way. Much of the constellation of *Hercules* is clear of the northern horizon, together with some of the northernmost stars in *Boötes*.

Ursa Major is climbing in the northeast, and all the far-flung outlying stars are clearly visible. Below the 'tail' is the inconspicuous constellation of *Canes Venatici*, and some observers at high latitudes may even be able to glimpse some of the stars of *Coma Berenices*, low on the horizon in the northeast. Between Ursa Major and the Milky Way, the whole of the constellations of *Draco* and *Ursa Minor* are clearly visible as is Cepheus beyond them. The chain of stars forming the faint constellation of *Lynx* lies between Ursa Major and the bright stars of Gemini.

This time of the year, the constellation of Perseus, with the famous variable star Algol and the Double Cluster, is at the meridian.

D

December – Looking South

Orion has now risen well clear of the horizon and is in the northeast. Above it are **Taurus** and **Auriga**, with **Gemini** farther to the east. Within the Milky Way to the east of Orion is the rather undistinguished constellation of **Monoceros**, which is without any distinct, bright stars. Below Orion is the small constellation of **Lepus**, and to its east, **Canis Major**, with brilliant **Sirius**, the brightest star in the sky (mag. -1.4). The long, winding constellation of **Eridanus** begins near **Rigel** in Orion and runs south, partly enclosing **Fornax** (which was once part of the larger constellation), until it ends at **Achernar** (α Eridani). Between Eridanus and Canis Major is the tiny, faint constellation of **Caelum** and the larger and brighter **Columba**.

South of Canis Major is the constellation of **Puppis**, once (with **Vela** and **Carina**) part of the large obsolete constellation of Argo Navis. **Canopus** (α Carinae) and Achernar are close to the horizon for observers at 30°N. West of Achernar is **Phoenix** and even farther west, below **Aquarius**, the constellation of **Piscis Austrinus** is slowly descending in the west. Between Piscis Austrinus in the west and Carina in the east lie several small constellations, the most conspicuous of which is **Grus**. There are also **Tucana** and **Hydrus**, with the **Small Magellanic Cloud** (SMC), then the **Large Magellanic Cloud** (LMC) in **Mensa** and **Dorado**, with the tiny constellation of **Reticulum** north of them. North of that again is the faint constellation of **Horologium**. Between Mensa and the Milky Way in Carina is the tiny constellation of **Volans**.

More faint constellations lie around the South Celestial Pole in **Octans**, notably **Chamaeleon** and **Apus**. To the west lies the larger **Pavo** and the rather undistinguished constellation of **Indus**. **Crux** has now swung round into the southeast. Both it and the small neighbouring constellation of **Musca** lie in the Milky Way. **Rigil Kentaurus** and **Hadar**, the two brightest stars in **Centaurus** are slightly farther south, with **Circinus** and **Triangulum Australe** between them and Pavo. Those in the far south are able to see all the stars in Centaurus and **Lupus**, together with the southernmost stars of **Scorpius** (the 'sting') and **Sagittarius**, as well as other constellations, such as **Norma**, **Ara**, **Telescopium** and **Corona Australis**.

D

Dark Sky
Sites

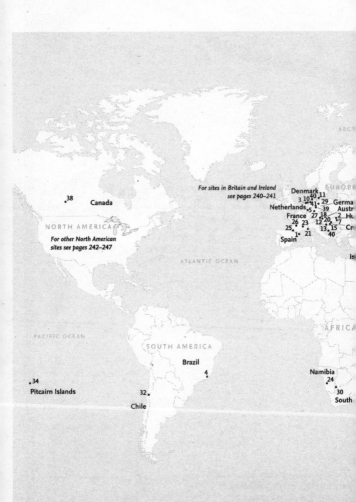

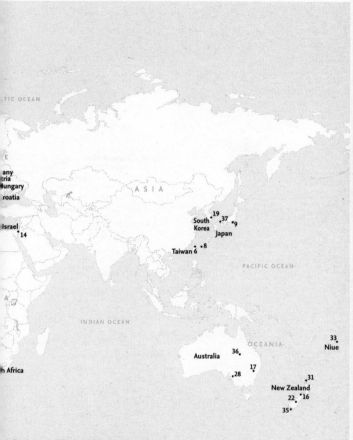

The 'rest of the world' site details
for numbers on this map are
given on pages 248–249.

Dark Sky Sites, Britain & Ireland

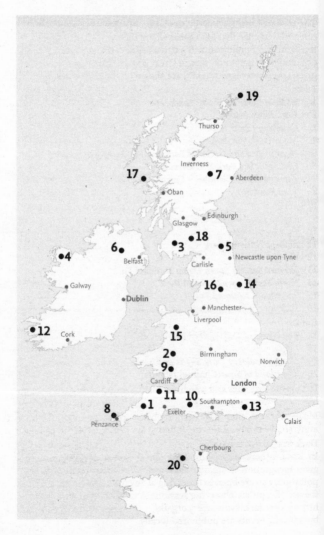

The International Dark-Sky Association (IDA) recognizes various categories of sites that offer areas where the sky is dark at night, free from light pollution and particularly suitable for astronomical observing. A number of sites in Great Britain and Ireland have been given specific recognition and are shown on the map. These are:

Parks

1   *Bodmin Moor Dark Sky Landscape*
2   *Elan Valley Estate*
3   *Galloway Forest Park*
4   *Mayo Dark Sky Park*
5   *Northumberland National Park and Kielder Water & Forest Park*
6   *OM Dark Sky Park & Observatory*
7   *Tomintoul and Glenlivet – Cairngorms*
8   *West Penwith*

Reserves

9   *Brecon Beacons National Park*
10   *Cranborne Chase*
11   *Exmoor National Park*
12   *Kerry*
13   *Moore's Reserve South Down National Park*
14   *North York Moors National Park*
15   *Snowdonia National Park*
16   *Yorkshire Dales National Park*

Communities

17   *The island of Coll* (Inner Hebrides, Scotland)
18   *Moffat*
19   *North Ronaldsay Dark Sky Island* (Orkney Islands, Scotland)
20   *The island of Sark* (Channel Islands)

Details of these sites and web links may be found at the IDA website: https://www.darksky.org/

Many of these sites have major observatories or other facilities available for public observing (often at specific dates or times).

Dark Sky Discovery Sites

In Britain there is also the **Dark Sky Discovery** organisation. This gives recognition to smaller sites, again free from immediate light pollution, that are open to observing at any time. Some sites are used for specific, public observing sessions. A full listing of sites is at: https://www.darkskydiscovery.org.uk/
but specific events are publicized locally.

US Dark Sky Sites

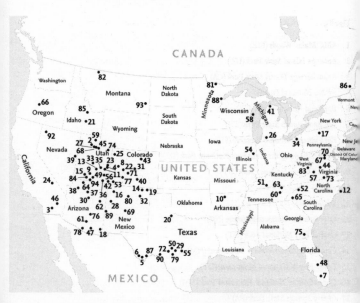

International Dark Sky Association Sites

The International Dark-Sky Association (IDA) recognizes various categories of sites that offer areas where the sky is dark at night, free from light pollution and particularly suitable for astronomical observing. There are numerous sites in North America, shown on the map and listed here.
Details of IDA are at: https://www.darksky.org/.

Information on the various categories and individual sites are at: https://www.darksky.org/our-work/conservation/idsp/

Many of these sites have major observatories or other facilities available for public observing (often at specific dates or times).

Parks

1 *AMC Maine Woods* (ME)

2 *Antelope Island State Park* (UT)

3 *Anza-Borrego Desert State Park* (CA)

4 *Arches National Park* (UT)

5 *Big Bend National Park* (TX)

6 *Big Bend Ranch State Park* (TX)

7 *Big Cypress National Preserve* (FL)

8 *Black Canyon of the Gunnison National Park* (CO)

9 *Bryce Canyon National Park* (UT)

10 *Buffalo National River* (AR)

11 *Canyonlands National Park* (UT)

12 *Cape Lookout National Seashore* (NC)

13 *Capitol Reef National Park* (UT)

14 *Capulin Volcano National Monument* (NM)

15 *Cedar Breaks National Monument* (UT)

16 *Chaco Culture National Historical Park* (NM)

17 *Cherry Springs State Park* (PA)

18 *Chiricahua National Monument* (AZ)

19 *Clayton Lake State Park* (NM)

20 *Copper Breaks State Park* (TX)

21 *Craters Of The Moon National Monument* (ID)

22 *Curecanti National Recreation Area* (CO)

23 *Dead Horse Point State Park* (UT)

24 *Death Valley National Park* (CA)

25 *Dinosaur National Monument* (CO)

26 *Dr. T.K. Lawless County Park* (MI)

27 *East Canyon State Park* (UT)

28 *El Morro National Monument* (NM)

29 *Enchanted Rock State Natural Area* (TX)

30 *Flagstaff Area National Monuments* (AZ)

31 *Florissant Fossil Beds National Monument* (CO)

65 *Pisgah Astronomical Research Institute* (NC)

66 *Prineville Reservoir State Park* (OR)

67 *Rappahannock County Park* (VA)

68 *Rockport State Park* (UT)

69 *Salinas Pueblo Missions National Monument* (NM)

70 *Sky Meadows State Park* (VA)

71 *Slumgullion Center* (CO)

72 *South Llano River State Park* (TX)

73 *Staunton River State Park* (VA)

74 *Steinaker State Park* (UT)

75 *Stephen C. Foster State Park* (GA)

76 *Tonto National Monument* (AZ)

77 *Top of the Pines* (CO)

78 *Tumacácori National Historical Park* (AZ)

79 *UBarU Camp and Retreat Center* (TX)

80 *Valles Caldera National Preserve* (NM)

81 *Voyageurs National Park* (MN)

82 *Waterton-Glacier International Peace Park* (Canada/MT)

83 *Watoga State Park* (WV)

84 *Zion National Park* (UT)

Reserves

85 *Central Idaho* (ID)

86 *Mont-Mégantic* (Québec)

Sanctuaries

87 *Black Gap Wildlife Management Area* (TX)

88 *Boundary Craters Canoe Area Wilderness* (MN)

89 *Cosmic Campground* (NM)

90 *Devils River State Natural Area – Del Norte Unit* (TX)

91 *Katahdin Woods and Waters National Monument* (ME)

92 *Massacre Rim* (NV)

93 *Medicine Rocks State Park* (MT)

94 *Rainbow Bridge National Monument* (UT)

RASC Recognized Dark-Sky Sites

Canadian Dark-Sky Sites
The Royal Astronomical Society of Canada (RASC) has developed formal guidelines and requirements for three types of light-restricted protected areas: Dark-Sky Preserves, Urban Star Parks and Nocturnal Preserves. The focus of the Canadian Program is primarily to protect the nocturnal environment; therefore, the outdoor lighting requirements are the most stringent, but also the most effective. Canadian Parks and other areas that meet these guidelines and successfully apply for one of these designations are officially recognized. Many parks across Canada have been designated in recent years – see the list below and the RASC website: https://www.rasc.ca/dark-sky-site-designations.

Dark-Sky Preserves

1   *Torrance Barrens Dark-Sky Preserve* (ON)

2   *McDonald Park Dark-Sky Park* (BC)

3   *Cypress Hills Inter-Provincial Park Dark-Sky Preserve* (SK/AB)

4   *Point Pelee National Park* (ON)

5   *Beaver Hills and Elk Island National Park* (AB)

6   *Mont-Mégantic International Dark-Sky Preserve* (QC)

7   *Gordon's Park* (ON)

8   *Grasslands National Park* (SK)

9   *Bruce Peninsula National Park* (ON)

10 *Kouchibouguac National Park* (NB)

11 *Mount Carleton Provincial Park* (NB)

12 *Kejimkujik National Park* (NS)

13 *Fundy National Park* (NB)

14 *Jasper National Park Dark-Sky Preserve* (AB)

15 *Bluewater Outdoor Education Centre – Wiarton* (ON)

16 *Wood Buffalo National Park* (AB)

17 *North Frontenac Township* (ON)

18 *Lakeland Provincial Park and Provincial Recreation Area* (AB)

19 *Killarney Provincial Park* (ON)

20 *Terra Nova National Park* (NL)

21 *Au Diable Vert* (QC)

22 *Lake Superior Provincial Park* (ON)

Urban Star Parks

23 *Irving Nature Park* (NB)

24 *Cattle Point, Victoria* (BC)

Nocturnal Preserves

25 *Ann and Sandy Cross Conservation Area* (AB)

26 *Old Man on His Back Ranch* (SK)

Dark Sky Parks – World

1 *Albanyà* (Spain)
2 *Bükk National Park* (Hungary)
3 *De Boschplatt* (Netherlands)
4 *Desengano State Park* (Brazil)
5 *Eifel National Park* (Germany)
6 *Hehuan Mountain* (Taiwan)
7 *Hortobágy National Park* (Hungary)
8 *Iriomote-Ishigaki National Park* (Japan)
9 *Kozushima Island* (Japan)
10 *Lauwersmeer National Park* (Netherlands)
11 *Møn and Nyord* (Denmark)
12 *Naturpark Attersee-Traunsee* (Austria)
13 *Petrova gora-Biljeg* (Croatia)
14 *Ramon Crater* (Israel)
15 *Vrani kamen* (Croatia)
16 *Wai-Iti* (New Zealand)
17 *Warrumbungle National Park* (Australia)
18 *Winklmoosalm* (Germany)
19 *Yeongyang Firefl y Eco Park* (South Korea)
20 *Zselic National Landscape Protection Area* (Hungary)

Dark Sky Reserves

21 *Alpes Azur Mercantour* (France)

22 *Aoraki Mackenzie* (New Zealand)

23 *Cévennes National Park* (France)

24 *NambiRand Nature Reserve* (Namibia)

25 *Pic du Midi* (France)

26 *Regional Natural Park of Millevaches in Limousin* (France)

27 *Rhön* (Germany)

28 *River Murray* (Australia)

29 *Westhavelland* (Germany)

Dark Sky Sanctuaries

30 *!Ae!Hai Kalahari Heritage Park* (South Africa)

31 *Aotea / Great Barrier Island* (New Zealand)

32 *Gabriela Mistral* (Chile)

33 *Niue* (New Zealand)

34 *Pitcairn Islands* (UK)

35 *Stewart Island / Rakiura* (New Zealand)

36 *The Jump-Up* (Australia)

Dark Sky Communities

37 *Bisei Town, Ibara City* (Japan)

38 *Bon Accord* (Canada)

39 *Fulda* (Germany)

40 *Pellworm Star Island* (Germany)

41 *Spiekeroog Star Island* (Germany)

Twilight Diagrams

Sunrise, sunset, twilight

For each individual month, we give details of sunrise and sunset times for nine cities across the world. But observing the stars is also affected by twilight, and this varies considerably from place to place. During the summer, especially at high latitudes, twilight may persist throughout the night and make it difficult to see the faintest stars. Beyond the Arctic and Antarctic Circles, of course, the Sun does not set for 24 hours at least once during the summer (and rise for 24 hours at least once during the winter). Even when the Sun does dip below the horizon at high latitudes, bright twilight persists throughout the night, so observing the stars is impossible.

There are three recognized stages of twilight: civil twilight, when the Sun is less than 6° below the horizon; nautical twilight, when the Sun is between 6° and 12° below the horizon; and astronomical twilight, when the Sun is between 12° and 18° below the horizon. Full darkness occurs only when the Sun is more than 18° below the horizon. During nautical twilight, only the very brightest stars are visible. (These are the stars that were used for navigation, hence the name for this stage.) During astronomical twilight, the faintest stars visible to the naked eye may be seen directly overhead, but are lost at lower altitudes. They become visible only once it is fully dark. The diagrams show the duration of twilight at the various cities. Of the locations shown, during the summer months there is full darkness at most of the cities, but it never occurs during the summer at the latitude of London. Observing conditions are most favourable at somewhere like Nairobi, which is very close to the equator, so there is not only little twilight, and a long period of full darkness, but there are also only slight variations in timing and duration throughout the year.

The diagrams also show the times of New and Full Moon (black and white symbols, respectively). As may be seen, at most locations during the year roughly half of New and Full Moon phases may come during daylight. For this reason, the exact phase may be invisible at one location, but be clearly seen elsewhere. The exact times of the events are given in the diagrams for each individual month.

Buenos Aires, Argentina – Latitude 34.7°S – Longitude 58.5°W

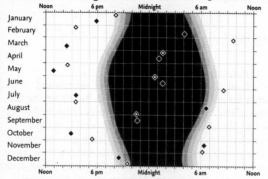

Cape Town, South Africa – Latitude 33.9°S – Longitude 18.5°E

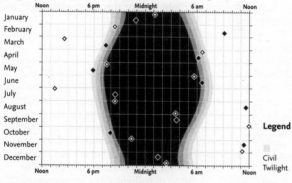

Legend

Civil Twilight

Nautical Twilight

Astronomical Twilight

Full Darkness

◇ Exact time of Full Moon

◆ Exact time of New Moon

London, UK – Latitude 51.5°N – Longitude 2.0°W

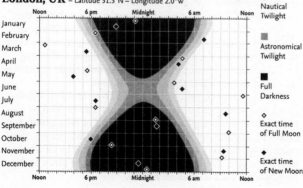

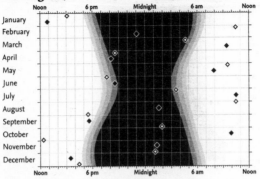

Los Angeles, USA – Latitude 34.0°N – Longitude 118.2°W

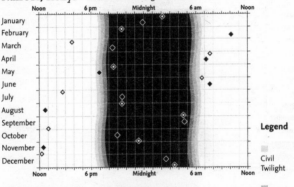

Nairobi, Kenya – Latitude 1.3°S – Longitude 36.8°E

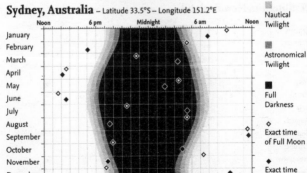

Sydney, Australia – Latitude 33.5°S – Longitude 151.2°E

Legend

Civil Twilight

Nautical Twilight

Astronomical Twilight

Full Darkness

◇ Exact time of Full Moon

◆ Exact time of New Moon

Tokyo, Japan — Latitude 35.7°N – Longitude 139.8°E

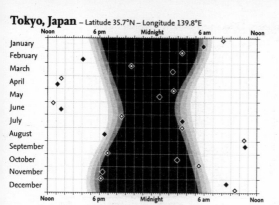

Washington, DC, USA — Latitude 38.9°N – Longitude 77.0°W

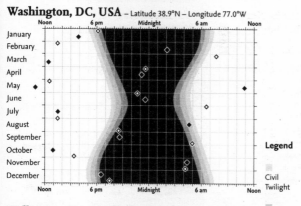

Wellington, New Zealand — Latitude 41.3°S – Longitude 174.8°E

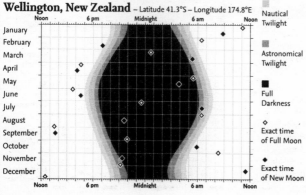

Legend

Civil Twilight

Nautical Twilight

Astronomical Twilight

Full Darkness

◇ Exact time of Full Moon

◆ Exact time of New Moon

Glossary and Tables

aphelion	The point on an orbit that is farthest from the Sun.
apogee	The point on its orbit at which the Moon is farthest from the Earth.
appulse	The apparently close approach of two celestial objects; two planets, or a planet and star.
astronomical unit	(AU) The mean distance of the Earth from the Sun, 149,597,870 km.
celestial equator	The great circle on the celestial sphere that is in the same plane as the Earth's equator.
celestial sphere	The apparent sphere surrounding the Earth on which all celestial bodies (stars, planets, etc.) seem to be located.
conjunction	The point in time when two celestial objects have the same celestial longitude. In the case of the Sun and a planet, superior conjunction occurs when the planet lies on the far side of the Sun (as seen from Earth). For Mercury and Venus, inferior conjunction occurs when they pass between the Sun and the Earth.
direct motion	Motion from west to east on the sky.
ecliptic	The apparent path of the Sun across the sky throughout the year. Also: the plane of the Earth's orbit in space.
elongation	The point at which an inferior planet has the greatest angular distance from the Sun, as seen from Earth.
equinox	The two points during the year when night and day have equal duration. Also: the points on the sky at which the ecliptic intersects the celestial equator. The vernal (northern spring) equinox is of particular importance in astronomy.
gibbous	The stage in the sequence of phases at which the illumination of a body lies between half and full. In the case of the Moon, the term is applied to phases between First Quarter and Full, and between Full and Last Quarter.

inferior planet	Either of the planets Mercury or Venus, which have orbits inside that of the Earth.
magnitude	The brightness of a star, planet or other celestial body. It is a logarithmic scale, where larger numbers indicate fainter brightness. A difference of 5 in magnitude indicates a difference of 100 in actual brightness, thus a first-magnitude star is 100 times as bright as one of sixth magnitude.
meridian	The great circle passing through the North and South Poles of a body and the observer's position; or the corresponding great circle on the celestial sphere that passes through the North and South Celestial Poles and also through the observer's zenith.
nadir	The point on the celestial sphere directly beneath the observer's feet, opposite the zenith.
occultation	The disappearance of one celestial body behind another, such as when stars or planets are hidden behind the Moon.
opposition	The point on a superior planet's orbit at which it is directly opposite the Sun in the sky.
perigee	The point on its orbit at which the Moon is closest to the Earth.
perihelion	The point on an orbit that is closest to the Sun.
retrograde motion	Motion from east to west on the sky.
superior planet	A planet that has an orbit outside that of the Earth.
vernal equinox	The point at which the Sun, in its apparent motion along the ecliptic, crosses the celestial equator from south to north. Also known as the First Point of Aries.
zenith	The point directly above the observer's head.
zodiac	A band, stretching 8° on either side of the ecliptic, within which the Moon and planets appear to move. It consists of 12 equal areas, originally named after the constellation that once lay within it.

The Constellations

There are 88 constellations covering the whole of the celestial sphere. The names themselves are expressed in Latin, and the names of stars are frequently given by Greek letters (see page 258) followed by the genitive of the constellation name or its three-letter abbreviation. The genitives, the official abbreviations and the English names of the various constellations are included.

Name	Genitive	Abbr.	English name
Andromeda	Andromedae	And	Andromeda
Antlia	Antliae	Ant	Air Pump
Apus	Apodis	Aps	Bird of Paradise
Aquarius	Aquarii	Aqr	Water Bearer
Aquila	Aquilae	Aql	Eagle
Ara	Arae	Ara	Altar
Aries	Arietis	Ari	Ram
Auriga	Aurigae	Aur	Charioteer
Boötes	Boötis	Boo	Herdsman
Caelum	Caeli	Cae	Burin
Camelopardalis	Camelopardalis	Cam	Giraffe
Cancer	Cancri	Cnc	Crab
Canes Venatici	Canum Venaticorum	CVn	Hunting Dogs
Canis Major	Canis Majoris	CMa	Big Dog
Canis Minor	Canis Minoris	CMi	Little Dog
Capricornus	Capricorni	Cap	Sea Goat
Carina	Carinae	Car	Keel
Cassiopeia	Cassiopeiae	Cas	Cassiopeia
Centaurus	Centauri	Cen	Centaur
Cepheus	Cephei	Cep	Cepheus
Cetus	Ceti	Cet	Whale
Chamaeleon	Chamaeleontis	Cha	Chameleon
Circinus	Circini	Cir	Compasses
Columba	Columbae	Col	Dove
Coma Berenices	Comae Berenices	Com	Berenice's Hair
Corona Australis	Coronae Australis	CrA	Southern Crown
Corona Borealis	Coronae Borealis	CrB	Northern Crown

Name	Genitive	Abbr.	English name
Corvus	Corvi	Crv	Crow
Crater	Crateris	Crt	Cup
Crux	Crucis	Cru	Southern Cross
Cygnus	Cygni	Cyg	Swan
Delphinus	Delphini	Del	Dolphin
Dorado	Doradus	Dor	Dorado
Draco	Draconis	Dra	Dragon
Equuleus	Equulei	Equ	Little Horse
Eridanus	Eridani	Eri	River Eridanus
Fornax	Fornacis	For	Furnace
Gemini	Geminorum	Gem	Twins
Grus	Gruis	Gru	Crane
Hercules	Herculis	Her	Hercules
Horologium	Horologii	Hor	Clock
Hydra	Hydrae	Hya	Water Snake
Hydrus	Hydri	Hyi	Lesser Water Snake
Indus	Indi	Ind	Indian
Lacerta	Lacertae	Lac	Lizard
Leo	Leonis	Leo	Lion
Leo Minor	Leonis Minoris	LMi	Little Lion
Lepus	Leporis	Lep	Hare
Libra	Librae	Lib	Scales
Lupus	Lupi	Lup	Wolf
Lynx	Lyncis	Lyn	Lynx
Lyra	Lyrae	Lyr	Lyre
Mensa	Mensae	Men	Table Mountain
Microscopium	Microscopii	Mic	Microscope
Monoceros	Monocerotis	Mon	Unicorn
Musca	Muscae	Mus	Fly
Norma	Normae	Nor	Set Square
Octans	Octantis	Oct	Octant
Ophiuchus	Ophiuchi	Oph	Serpent Bearer
Orion	Orionis	Ori	Orion
Pavo	Pavonis	Pav	Peacock
Pegasus	Pegasi	Peg	Pegasus
Perseus	Persei	Per	Perseus

Name	Genitive	Abbr.	English name
Phoenix	Phoenicis	Phe	Phoenix
Pictor	Pictoris	Pic	Painter's Easel
Pisces	Piscium	Psc	Fishes
Piscis Austrinus	Piscis Austrini	PsA	Southern Fish
Puppis	Puppis	Pup	Stern
Pyxis	Pyxidis	Pyx	Compass
Reticulum	Reticuli	Ret	Net
Sagitta	Sagittae	Sge	Arrow
Sagittarius	Sagittarii	Sgr	Archer
Scorpius	Scorpii	Sco	Scorpion
Sculptor	Sulptoris	Scu	Sculptor
Scutum	Scuti	Sct	Shield
Serpens	Serpentis	Ser	Serpent
Sextans	Sextantis	Sex	Sextant
Taurus	Tauri	Tau	Bull
Telescopium	Telescopii	Tel	Telescope
Triangulum	Trianguli	Tri	Triangle
Triangulum Australe	Trianguli Australis	TrA	Southern Triangle
Tucana	Tucanae	Tuc	Toucan
Ursa Major	Ursae Majoris	UMa	Great Bear
Ursa Minor	Ursae Minoris	UMi	Lesser Bear
Vela	Velorum	Vel	Sails
Virgo	Virginis	Vir	Virgin
Volans	Volantis	Vol	Flying Fish
Vulpecula	Vulpeculae	Vul	Fox

The Greek Alphabet

α	Alpha	ι	Iota	ρ	Rho
β	Beta	κ	Kappa	σ (ς)	Sigma
γ	Gamma	λ	Lambda	τ	Tau
δ	Delta	μ	Mu	υ	Upsilon
ε	Epsilon	ν	Nu	φ (φ)	Phi
ζ	Zeta	ξ	Xi	χ	Chi
η	Eta	o	Omicron	ψ	Psi
θ (ϑ)	Theta	π	Pi	ω	Omega

Asterisms

Apart from the constellations (88 of which cover the whole sky), listed on pages 256–258, certain groups of stars, which may form a small part of a larger constellation, are readily recognizable and have been given individual names. These groups are known as *asterisms*, and the most famous (and well-known) is the 'Plough' or 'Big Dipper', the common name for the seven brightest stars in the constellation of Ursa Major, the Great Bear. The names and details of some asterisms mentioned in this book are given in this list.

Some common asterisms

Belt of Orion	δ, ε and ζ Orionis
Big Dipper	α, β, γ, δ, ε, ζ and η Ursae Majoris
Cat's Eyes	λ and υ Scorpii
Circlet	γ, θ, ι, λ and κ Piscium
False Cross	ε and ι Carinae and δ and κ Velorum
Fish Hook	α, β, δ and π Scorpii
Guards (or Guardians)	β and γ Ursae Minoris
Head of Cetus	α, γ, ξ², μ and λ Ceti
Head of Draco	β, γ, ξ and ν Draconis
Head of Hydra	δ, ε, ζ, η, ρ and σ Hydrae
Job's Coffin	α, β, γ and δ Delphini
Keystone	ε, ζ, η and π Herculis
Kids	ε, ζ and η Aurigae
Little Dipper	β, γ, η, ζ, ε, δ and α Ursae Minoris
Lozenge	= Head of Draco
Milk Dipper	ζ, γ, σ, φ and λ Sagittarii
Plough	= Big Dipper
Pointers	α and β Ursae Majoris
Pot	= Saucepan
Saucepan	ι, θ, ζ, ε, δ and η Orionis
Sickle	α, η, γ, ζ, μ and ε Leonis
Southern Pointers	α and β Centauri
Square of Pegasus	α, β and γ Pegasi with α Andromedae
Sword of Orion	θ and ι Orionis
Teapot	γ, ε, δ, λ, φ, σ, τ and ζ Sagittarii
Wain (or Charles' Wain)	= Big Dipper
Water Jar	γ, η, κ and ζ Aquarii
Y of Aquarius	= Water Jar

107 Named stars brighter than magnitude 2.75

Name	Con	Mag	Name	Con	Mag
Achernar	α Eri	0.45	Aspidiske	ι Car	2.21
Acrab	β Sco	2.56	Athebyne	η Dra	2.73
Acrux	α Cru	0.77	Atria	α TrA	1.91
Adhara	ε CMa	1.50	Avior	ε Car	1.86
Aldebaran	α Tau	0.87	Bellatrix	γ Ori	1.64
Alderamin	α Cep	2.45	Betelgeuse	α Ori	0.45
Algieba	γ Leo	2.01	Canopus	α Car	-0.62
Algol	β Per	2.09	Capella	α Aur	0.08
Alhena	γ Gem	1.93	Caph	β Cas	2.28
Alioth	ε UMa	1.76	Castor	α Gem	1.58
Aljanah	ε Cyg	2.48	Deneb	α Cyg	1.25
Alkaid	η UMa	1.85	Denebola	β Leo	2.14
Almach	γ And	2.10	Diphda	β Cet	2.04
Alnair	α Gru	1.73	Dschubba	δ Sco	2.29
Alnilam	ε Ori	1.69	Dubhe	α UMa	1.81
Alnitak	ζ Ori	1.74	Elnath	β Tau	1.65
Alphard	α Hya	1.99	Eltanin	γ Dra	2.24
Alphecca	α CrB	2.22	Enif	ε Peg	2.38
Alpheratz	α And	2.07	Fomalhaut	α PsA	1.17
Alsephina	δ Vel	1.93	Gacrux	γ Cru	1.59
Altair	α Aql	0.76	Gienah	γ Crv	2.58
Aludra	η CMa	2.45	Hadar	β Cen	0.61
Ankaa	α Phe	2.40	Hamal	α Ari	2.01
Antares	α Sco	1.06	Hassaleh	ι Aur	2.69
Arcturus	α Boo	-0.05	Izar	ε Boo	2.35
Arneb	α Lep	2.58	Kaus Australis	ε Sgr	1.79
Ascella	ζ Sgr	2.60	Kaus Media	δ Sgr	2.72

Name	Con	Mag	Name	Con	Mag
Kochab	β UMi	2.07	**Procyon**	α CMi	0.40
Kraz	β Crv	2.65	**Rasalhague**	α Oph	2.08
Larawag	ε Sco	2.29	**Regulus**	α Leo	1.36
Lesath	υ Sco	2.70	**Rigel**	β Ori	0.18
Mahasim	ϑ Aur	2.65	**Rigil Kentaurus**	α Cen	-0.29
Markab	α Peg	2.49	**Ruchbah**	δ Cas	2.66
Markeb	κ Vel	2.47	**Sabik**	η Oph	2.43
Menkalinan	β Aur	1.90	**Sadr**	γ Cyg	2.23
Menkar	α Cet	2.54	**Saiph**	κ Ori	2.07
Menkent	ϑ Cen	2.06	**Sargas**	ϑ Sco	1.86
Merak	β UMa	2.34	**Scheat**	β Peg	2.44
Miaplacidus	β Car	1.67	**Schedar**	α Cas	2.24
Mimosa	β Cru	1.25	**Shaula**	λ Sco	1.62
Mintaka	δ Ori	2.25	**Sheratan**	β Ari	2.64
Mirach	β And	2.07	**Sirius**	α CMa	-1.44
Mirfak	α Per	1.79	**Spica**	α Vir	0.98
Mirzam	β CMa	1.98	**Suhail**	λ Vel	2.23
Mizar	ζ UMa	2.23	**Tarazed**	γ Aql	2.72
Muphrid	η Boo	2.68	**Tiaki**	β Gru	2.07
Naos	ζ Pup	2.21	**Unukalhai**	α Ser	2.63
Nunki	σ Sgr	2.05	**Vega**	α Lyr	0.03
Peacock	α Pav	1.94	**Wezen**	δ CMa	1.83
Phact	α Col	2.65	**Yed Prior**	δ Oph	2.73
Phecda	γ UMa	2.41	**Zosma**	δ Leo	2.56
Polaris	α UMi	1.97	**Zubenelgenubi**	α Lib	2.75
Pollux	β Gem	1.16	**Zubeneschamali**	β Lib	2.61
Porrima	γ Vir	2.74			

Further Information

Books

Bone, Neil (1993), *Observer's Handbook: Meteors*, George Philip, London & Sky Publishing Corp., Cambridge, Mass.

Cook, J., ed. (1999), *The Hatfield Photographic Lunar Atlas*, Springer-Verlag, New York

Dunlop, Storm (2006), *Wild Guide to the Night Sky*, Harper Perennial, New York & Smithsonian Press, Washington DC

Dunlop, Storm (2012), *Practical Astronomy*, 2nd edn, Firefly, Buffalo

Dunlop, Storm, Rükl, Antonin & Tirion, Wil (2005), *Collins Atlas of the Night Sky*, HarperCollins, London & Smithsonian Press, Washington DC

Grego, Peter (2016), *Moon Observer's Guide*, Firefly, Richmond Hill

Heifetz, Milton D. & Tirion, Wil (2017), *A Walk through the Heavens*, 4th edn, Cambridge University Press, Cambridge

Mellinger, Axel & Hoffmann, Susanne (2005), *The New Atlas of the Stars*, Firefly, Richmond Hill

O'Meara, Stephen J. (2008), *Observing the Night Sky with Binoculars*, Cambridge University Press, Cambridge

Pasachoff, Jay M. (1999), *Peterson Field Guides: Stars and Planets*, 4th edn., Houghton Mifflin, Boston

Ridpath, Ian, ed. (2003), *Oxford Dictionary of Astronomy*, 2nd edn, Oxford University Press, Oxford & New York

Ridpath, Ian, ed. (2004), *Norton's Star Atlas*, 20th edn, Pi Press, New York

Ridpath, Ian (2018), *Star Tales*, 2nd edn, Lutterworth Press, Cambridge UK

Ridpath, Ian & Tirion, Wil (2004), *Collins Gem – Stars*, HarperCollins, London

Ridpath, Ian & Tirion, Wil (2017), *Collins Pocket Guide Stars and Planets*, 5th edn, HarperCollins, London

Ridpath, Ian & Tirion, Wil (2012), *Monthly Sky Guide*, 10th edn, Dover Publications, New York

Rükl, Antonín (1990), *Hamlyn Atlas of the Moon*, Hamlyn, London & Astro Media Inc., Milwaukee

Rükl, Antonín (2004), *Atlas of the Moon*, Sky Publishing Corp., Cambridge, Mass.

Scagell, Robin (2014), *Stargazing with a Telescope*, Firefly, Richmond Hill

Scagell, Robin (2015), *Firefly Complete Guide to Stargazing*, Firefly, Richmond Hill

Scagell, Robin & Frydman, David (2014), *Stargazing with Binoculars*, Firefly, Richmond Hill

Sky & Telescope (2017), *Astronomy 2018*, Sky Publishing Corp., Cambridge, Mass.

Stimac, Valerie (2019), *Dark Skies: A Practical Guide to Astrotourism*, Lonely Planet

Tirion, Wil (2011), *Cambridge Star Atlas*, 4th edn, Cambridge University Press, Cambridge

Tirion, Wil & Sinnott, Roger (1999), *Sky Atlas 2000.0*, 2nd edn, Sky Publishing Corp., Cambridge, Mass. & Cambridge University Press, Cambridge

Journals

Astronomy, Astro Media Corp., 21027 Crossroads Circle, P.O. Box 1612, Waukesha, WI 53187-1612.
http://www.astronomy.com

Sky & Telescope, Sky Publishing Corp., Cambridge, MA 02138-1200.
http://www.skyandtelescope.com/

Societies

American Association of Variable Star Observers (AAVSO), 49 Bay State Rd., Cambridge, MA 02138. Although primarily concerned with variable stars, the AAVSO also has a solar section.

American Astronomical Society (AAS), 1667 K Street NW, Suite 800, Washington, DC 20006, New York.
http://aas.org/

American Meteor Society (AMS), Geneseo, New York.
http://www.amsmeteors.org/

Association of Lunar and Planetary Observers (ALPO), ALPO Membership Secretary/Treasurer, P.O. Box 13456, Springfield, IL 62791-3456. An organization concerned with all forms of amateur astronomical observation, not just the Moon and planets, with numerous coordinated observing sections.
http://alpo-astronomy.org/

Astronomical League (AL), 9201 Ward Parkway Suite #100,
Kansas City, MO 64114.
An umbrella organization consisting of over 240 local amateur
astronomical societies across the United States.
https://www.astroleague.org/

British Astronomical Association (BAA), Burlington House, Piccadilly,
London W1J 0DU.
The principal British organization (but with a worldwide
membership) for amateur astronomers (with some professional
members), particularly for those interested in carrying out
observational programs.
http://www.britastro.org/

International Meteor Organization (IMO)
An organization coordinating observations of meteors worldwide.
http://www.imo.net/

Royal Astronomical Society of Canada (RASC), 203 – 4920 Dundas Street W.,
Toronto, ON M9A 1B7.
The principal Canadian astronomical organization, with both
professional and amateur members. It has 28 local centres.
http://rasc.ca/

Software

Planetary, Stellar and Lunar Visibility (planetary and eclipse freeware):
Alcyone Software, Germany.
http://www.alcyone.de

Redshift, Redshift-Live.
http://www.redshift-live.com/en/

Starry Night & Starry Night Pro, Sienna Software Inc., Toronto, Canada.
http://www.starrynight.com

Internet sources

There are numerous sites about all aspects of astronomy, and all have
numerous links. Although many amateur sites are excellent, treat any
statements and data with caution. The sites listed below offer accurate
information. Please note that the URLs may change. If so, use a good search
engine, such as Google, to locate the information source.

Information

Astronomical data (inc. eclipses) HM Nautical Almanac Office:
http://astro.ukho.gov.uk

Auroral information Michigan Tech:
http://www.geo.mtu.edu/weather/aurora/

Comets JPL Solar System Dynamics:
http://ssd.jpl.nasa.gov/

Deep-sky objects Saguaro Astronomy Club Database:
http://www.virtualcolony.com/sac/

Eclipses NASA Eclipse Page:
http://eclipse.gsfc.nasa.gov/eclipse.html

Moon (inc. Atlas) Inconstant Moon:
http://www.inconstantmoon.com/

Planets Planetary Fact Sheets:
http://nssdc.gsfc.nasa.gov/planetary/planetfact.html

Satellites (inc. International Space Station)
Heavens Above: http://www.heavens-above.com/
Visual Satellite Observer: http://www.satobs.org/

Star Chart
http://www.skyandtelescope.com/observing/interactive-sky-watching-tools/interactive-sky-chart/

What's Visible
Skyhound: http://www.skyhound.com/sh/skyhound.html
Skyview Cafe: http://www.skyviewcafe.com

Institutes and Organizations

European Space Agency: http://www.esa.int/

International Dark-Sky Association: http://www.darksky.org/

RASC Dark Sky: https://www.rasc.ca/dark-sky-site-designations

Jet Propulsion Laboratory: http://www.jpl.nasa.gov/

Lunar and Planetary Institute: http://www.lpi.usra.edu/

National Aeronautics and Space Administration: http://www.hq.nasa.gov/

Solar Data Analysis Center: http://umbra.gsfc.nasa.gov/

Space Telescope Science Institute: http://www.stsci.edu/

Acknowledgements

23	NASA/GSFC/SVS
48	Stephen Pitt
57	NASA
58	Wikimedia Commons
67	Svend Buhl / Meteorite Recon CC by SA 3.0
75	Natural History Museum, London
80	NASA
85	NASA
99	NASA
100 (both)	NASA
108	Peter Komka/EPA-EFE/Shutterstock
115 (both)	Wikimedia
122	Steve Edberg
131	Gestrgangleri
139	Wil Tirion
147	JAXA/ISAS/DARTS/Kevin M. Gill
148	NASA
164	European Southern Observatory (ESO)
165	GAIA/European Space Administration (ESA)
173	H. Raab (Wikipedia)
174	Royal Canadian Museum
178	NASA/Joel Kowsky
179	NASA/Dylan O'Donnell
183 (both)	NASA
204	NASA
205	NASA
206	NASA
220	NASA
227	Sjbmgrtl

The authors would also like to thank Barry Hetherington.

Specialist editorial support was provided by Tania de Sales Marques, Planetarium Astronomer and Jacob Foster, Public Astronomy Officer at the Royal Observatory, part of Royal Museums Greenwich.

Index

Other titles by Storm Dunlop and Wil Tirion

2023 Guide to the Night Sky: Britain and Ireland
978-0-00-839354-0

2023 Guide to the Night Sky: North America
978-0-00-853258-1

2023 Guide to the Night Sky: Southern Hemisphere
978-0-00-853257-4

Latest editions of our bestselling month-by-month guides for exploring the skies. These guides are an easy introduction to astronomy and a useful reference for seasoned stargazers.

Collins Planisphere | 978-0-00-754075-4

Easy-to-use practical tool to help astronomers to identify the constellations and stars every day of the year. For latitude 50°N, suitable for use anywhere in Britain and Ireland, Northern Europe, Canada and Northern USA.

Also available

Astronomy Photographer of the Year: Collection 11
978-0-00-853262-8

Winning and shortlisted images from the 2022 Astronomy Photographer of the Year competition, hosted by the Royal Observatory, Greenwich. The images include aurorae, galaxies, our Moon, our Sun, people and space, planets, comets and asteroids, skyscapes, stars and nebulae.

Stargazing | 978-0-00-819627-1

The prefect manual for beginners to astronomy – introducing the world of telescopes, planets, stars, dark skies and celestial maps.

Moongazing | 978-0-00-830500-0

An in-depth guide for all aspiring astronomers and Moon observers, with detailed Moon maps. Covers the history of lunar exploration and the properties of the Moon, its origin and orbit.

Northern Lights | 978-0-00-846555-1

Discover the incomparable beauty of the Northern Lights with this accessible guide for both aspiring astronomers and seasoned night sky observers alike.

Observing our Solar System | 978-0-00-853261-1

Study the ever-changing face of the Moon, watch the steady march of the planets against the stars, witness the thrill of a meteor shower, or the memory of a once-in-a-generation comet.